How to Solve MATH PUZZLES

HOW TO SOLVE MATH PUZZLES

Anu Sehgal

Lotus PRESS

4735/22, Prakash Deep Building,
Ansari Road, Darya Ganj,
New Delhi- 110002

LOTUS PRESS Publishers & Distributors
Unit No. 220, 2nd Floor, 4735/22, Prakash Deep Building,
Ansari Road, Darya Ganj, New Delhi-110002
Ph.: 32903912, 23280047, 098118-38000
E-mail: lotuspress1984@gmail.com
Visit us: www.lotuspress.co.in

How to Solve Math Puzzles

ISBN 978-81-8382-319-7

Published by: **Lotus Press Publishers & Distributors**, New Delhi-110002.
Printed at: Bharat Offset Works, Delhi.

PREFACE

Mathematics is a formal area of teaching and learning which was developed about 5,000 years ago by the Sumerians. They did this at the same time as they developed reading and writing. However, the roots of mathematics go back much more than 5,000 years.

Mathematics tends to be defined by the types of problems it addresses, the methods it uses to address these problems, and the results it has achieved. It is divides it into the following three categories:

Mathematics as a human endeavor.

For example, consider the math of measurement of time such as years, seasons, months, weeks, days, and so on. Or, consider the measurement of distance, and the different systems of distance measurement that developed throughout the world. Or, think about math in art, dance, and music. There is a rich history of human development of mathematics and mathematical uses in our present day society.

Mathematics as a discipline.

Mathematics is a broad and deep subject that is growing

day by day. Nowadays, the Ph.D. researchers are focused on definitions, theorems and proofs related to a single problem in mathematics.

Mathematics as an interdisciplinary language and tool

Like reading and writing, math is an important component of learning and "doing" in each academic discipline. Mathematics is such a useful language and tool that it is considered one of the "basics" in our formal educational system.

The approach taken in this book – called (book name) – seeks to address this deficiency by teaching us how to frame and solve "unstructured" problems. Our book contains hard-to-solve maths puzzle, Sudoku and some brain teasers. Based on a variety of puzzles that are inherently unstructured, this book increases our mathematical awareness and ability to solve difficult problems.

We hope the book will help many readers to known the facts of history. Please be free to communicate with any feedback or suggestions for the next editions.

Author

CONTENTS

Chapter 1

Warm Up

1. Which number replaces the question marks?

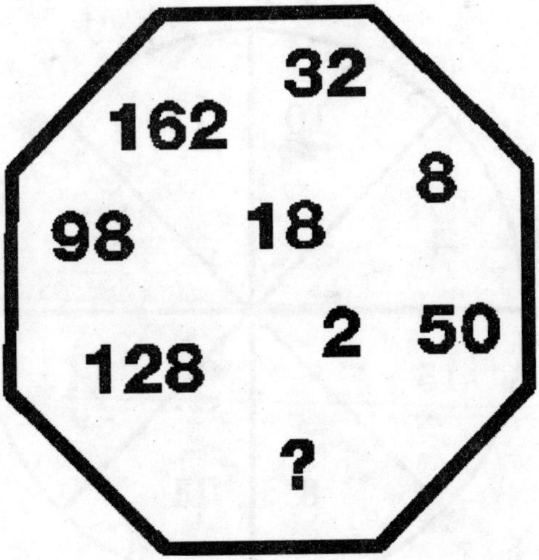

Answer

72

Moving a clockwise, numbers are double the square of the first 9 numbers.

2. What is the next number in the sequence below?

1 4 9 16 25 36 ?

Answer

The answer is 49. The difference between each two consecutive numbers grows in the order of 3, 5, 7, 9, 11, and 13 . All odd numbers.

3. What is the missing number in the pie below?

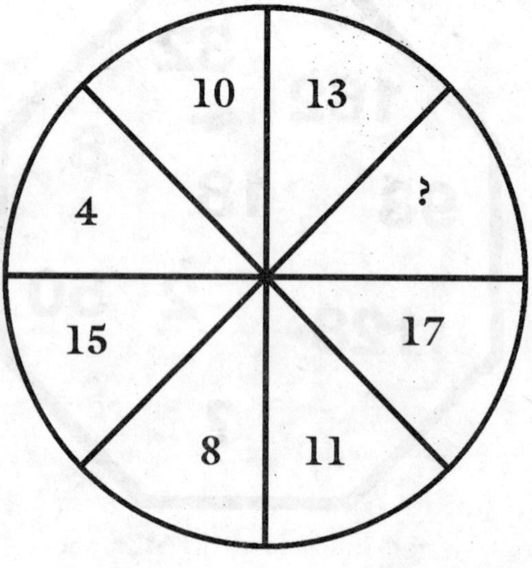

Answer

The missing number is 6.The two numbers opposite each other always total same i.e. 21.

4. Can you determine the missing number in the box? The same rule of logic applies to all three boxes?

4	5	6
8	10	12
16	20	?
32	40	48

Answer

The missing number is 24.

The rule is to double the first number in each rectangle and place that number in the next box below it.

5. In a group of 28 junior high school students, 7 take French, 10 take Spanish, and 4 take both languages. The students taking both French and Spanish are not counted with the 7 taking French or the 10 taking Spanish. How many students are not taking either French or Spanish?

Answer

Seven students are not taking a language. Add 7 + 10 +4 to get 21.

Then subtract 21 from the total students: 28 - 21 = 7.

6. What number should replace the question mark?

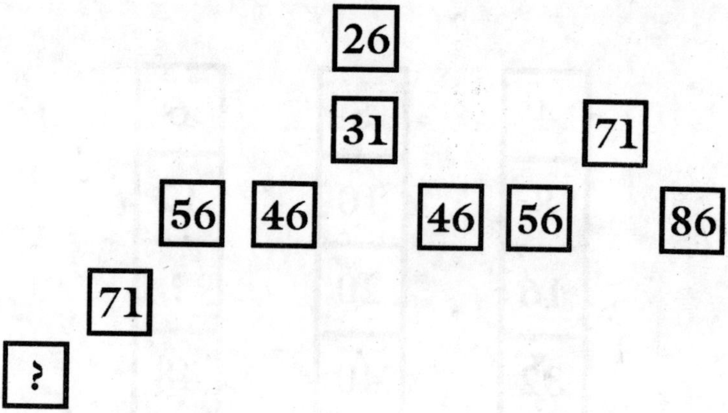

Answer

86

Add five straightdown; Add ten sideways; Add fifteen diagonally.

7. Which number continues this sequence?

18	20	24	32	?

Answer

48

Numbers advance by 2, 4, 8 and 16.

8. Which number does not fit in this sequence?

1 - 2 - 3 - 4 - 5 - 6 - 7 - 8 - 14 - 15 - 30

Answer

14

The sequence follows the formula double the previous number then add 1. etc.

9. Which number is missing from the box?

8	3	4
1	5	?
6	7	2

Answer

9

Each row; column and diagonal adds up to 15.

10. Which number completes this sequence?

123 117 108 99 ?

Answer

81

As you go down, subtract the sum of the separate digits in each number from itself to give the next number.

11. Which number is missing from this wheel?

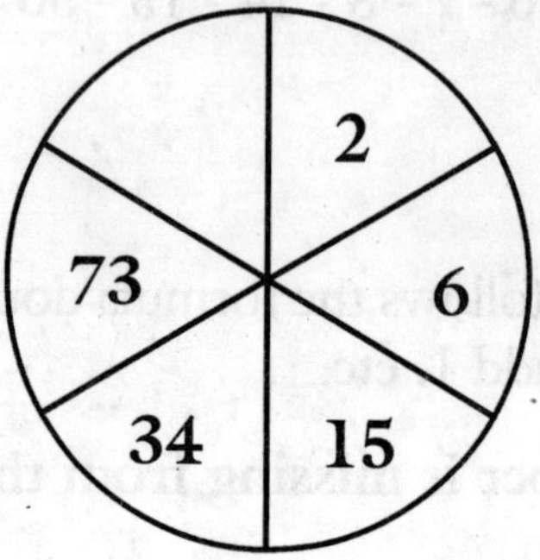

Answer

152

As you move clockwise multipley the previous number by 2 and add 2, 3, 4, 5 and 6 respectively.

12. What is X?

24 81 63 26 412 8 25 X

Answer

6

The series is spaced incorrectly. When the spacing is correct it becomes: 2 4 8 16 32 64 128, 256, which is an obvious doubling-up series..

13. What number goes into the empty brackets?

16	(4	2	5	6)
9	(3	8	1)	
25	(	)		

Answer

5 6 2 5

The first number inside the brackets is the square root of the number outside the brackets. The remaining number inside the brackets is the square of the number outside the brackets.

14. What comes next in this series?

1 2 6 24 120 720 –

Answer

5040

Multiply each number by 2, 3, 4, 5 and 6 and (finally) 7.

15. What comes next in the series?

16 72 38 94 50 –

Answer

16

Each number reverses the previous number and adds 1 to each digit. Thus, in the first two terms, 16 reversed is 61, which then changes to 72. In the penultimate term, 50 reversed becomes 05, which in turn becomes 16 - by adding 1 to each digit.

16. Which number goes in the center of the third triangle?

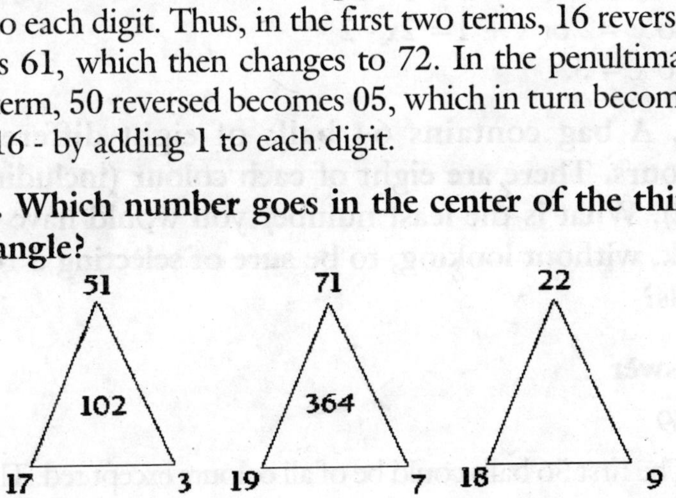

Answer

36

The answer in the centre of each triangle equals the difference between the top and left hand values, multiplied by the right hand value.

17. If I had one more sister I would have twice as many sisters as brothers. If I had one more brother I would have the same number of each. How many brothers and sisters have I?

Answer

Three sisters and two brothers.

This can be solved by simple deduction, but if algebra is used, let x be the number of sisters and y the number of brothers:

$x + 1 = 2y$
$y + 1 = x$
Therefore, $y + 1 + 1 = 2y$
so $y = 2$ or $x + 1 = 2x - 2$
so $x = 3$.

18. A bag contains 64 balls of eight different colours. There are eight of each colour (including red). What is the least number you would have to pick, without looking, to be sure of selecting 3 red balls?

Answer

59

The first 56 balls could be of all colours except red. This

would leave 8 balls, all of which are red. so any three chosen would be red.

19. Which number goes in the middle?

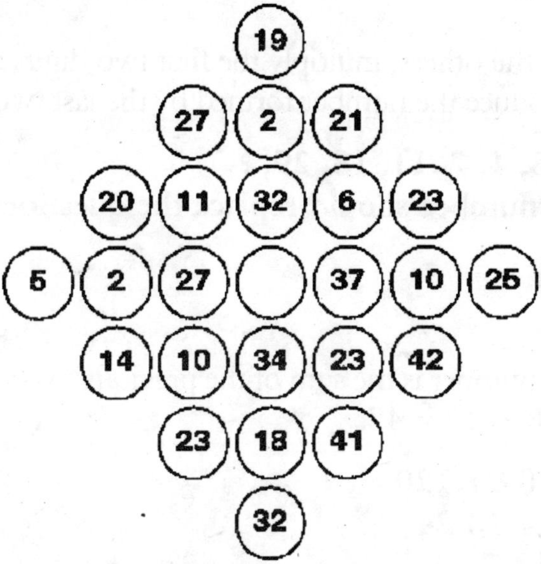

Answer

17

Moving diagonally downwards from left to right, numbers increase by the same amount each line.

20. 10, 50, 13, 45, 18, 38, ?, ?
What two numbers should replace the question marks?

Answer

25, 29.

There are two alternate sequences: start at 10 and add 3, 5, 7; start at 50 and deduct 5, 7, 9.

21. Which number is the odd one out?
9654 4832 5945 7642 7963 8216 3649

Answer

3649

In all the others, multiply the first two digits together to produce the number formed by the last two digits.

22. 1, 3, 4, 7, 11, 18, 29, ?
Which number should replace the question mark?

Answer

47

Each number is the sum of the previous two numbers, i.e. 18 + 29 = 47.

23.

70	**91**	**120**
14	**13**	**24**
5	**7**	**?**

What number should replace the question mark?

Answer

5

$70 \div 14 = 5$;

$91 \div 13 = 7$,

$120 \div 24 = 5$;

24. What number should replace the question mark?
123, 124, 126, 132, 133, 136, 142, 143, 147, ?

Answer

154

Add the first digit with 1 to arrive at the second number i.e. 123 + 1 = 124, then the second digit with 2 i.e. 124 + 2 = 126, then the third digit etc.

25. What number comes next?
482, 693, 714, 826, 937, ?

Answer

148

The numbers 48269371 are being repeated in the same sequence.

26. What numbers should replace the question marks?

2 6 30 ?
3 5 11 ?

Answer

330 and 41

Multiply the two numbers of the previous pair together (i.e. 2&3)to obtain the top number, and add the same two numbers to get the bottom number. So, 30 × 11 = 330, 30 + 11 = 41.

27. Which number is the odd one out?
159
248
963
357
951
852

Answer

248

In the rest there is the same difference between each digit, eg: 8 (–3) 5 (–3) 2.

28. What number continues the sequence? 25, 50, 27, 46, 31, 38, 39, ?

Answer

22

There are two alternate sequences, the first increases by 2, 4, 8 etc and the second decreases by 4, 8, 16 etc.

29. What number comes next in this sequence? 1, 3, 11, 47, ?

Answer

239

$1 \times 2 + 1 = 3$;

$3 \times 3 + 2 = 11$;

$11 \times 4 + 3 = 47$;

$47 \times 5 + 4 = 239$;

30.

83	**64**	**45**	**?**	**?**
96	**65**	**34**	**?**	**?**

What numbers should replace the question marks?

Answer

2 6, 0 3.

Looking across, the numbers in the same position in each pair progress: 8, 6, 4, 2; 3, 4, 5, 6; 9, 6, 3, 0; 6, 5, 4, 3.

31. Which number is the odd one out?
571219
461016
831114
461016
971613
781523

Answer

971613

Numbers are obtained by adding pairs of digits, ie with 571219.

5 + 7 = 12; 7 + 12 = 19.

To follow this same pattern 971613 would have to be 971623.

32. 12, 34, 10, 11, 12, 5, 6 ,7, 13, 14, ?, ?
Logically, what two numbers should replace the question marks?

Answer

15 and 11

1 + 2 + 3 + 4 = 10, then insert next two numbers 11, 12 and add digits

1 + 1 + 1 + 2 = 5.

Insert next two numbers 6, 7 and add digits 6 + 7 = 13. Insert next two numbers 14, 15 and add digits 1 + 4 + 1 + 5 = 11.

33. My watch was correct at noon, after which it started to lose 17 minutes per hour until six hours

ago it stopped completely. It now shows the time as 2.52 pm. What time is it now?

Answer

10 pm.
12 noon = 12 noon,
1 pm = 12.43,
2 pm = 1.26,
3 pm = 2.09,
4 pm = 2.52,+6 hours = 10 pm.

34. A green grocer received a boxful of tomatoes and on opening the box found that several had gone bad. He then counted them up so that he could make a formal complaint and found that 68 were mouldy, which was 16 per cent of the total contents of the box. How many tomatoes were in the box?

Answer

425. [(68 ÷ 16) × 100]

35. Midway through his round a golfer hits a magnificent 210 yard drive, which brings his average length per drive for the round up to now from 156 to 162 yards. How far would he have had to hit the drive to bring his average length of drive up from 156 to 165 yards?

Answer

237 yards
Eight holes average 156 = 1, 248 yards,
Nine holes average 162 = 1, 458 yards (+ 210),

Nine holes average 165 = 1, 485 yards (+ 237).

36. What numbers should go on the bottom line?

3	6	9	15
8	2	10	12
11	8	19	27
19	10	29	39
?	?	?	?

Answer

30 18 48 66.

Starting at the top, add pairs of numbers in each column to arrive at the next number.

37. What number continues the following sequence?

759, 675, 335, 165, ?

Answer

80

Each number is formed by multiplying the number formed by the first two digits of the previous number by its third digit. So, 16 × 5 = 80;

38. A statue is being carved by a sculptor. The original piece of marble weighed 250 kg. In the first week 30 per cent is cut away. In the second week 20 per cent of the remainder is cut away. In the third week the statue is completed when 25 percent of the remainder is cut away. What is the weight of the final statue?

Answer

105 kg

$= 250 \times 0.7 \times 0.8 \times 0.75.$

39. Which is the odd number out?

9421, 7532, 9854, 8612, 6531, 8541

Answer

8612.

All the other numbers have their digits in descending order.

40. What is the factorial of 5?

a. 1
b. 120
c. 60
d. 25
e. 15

Answer B

41. What is the value of 56 + –19?

Answer

37

The rule is to replace + – with –. Thus 56 – 19 = 37.

42. You have two bags each containing six balls. The first bag contains balls numbered 1 to 6 and the second bag contains balls numbered 7 to 12. A ball is drawn out of bag one and another ball is drawn out of bag two. What are the chances that

at least one of the balls drawn out is an odd numbered ball?

Answer

3 in 4.

The possible combinations of drawing the balls are
odd – odd
odd – even
even – odd
even – even

The chance of drawing at least one odd-numbered ball is therefore three in four.

43. 0, 1, 2, 4, 6, 9, 12, 16, ?

What number should replace the question mark?

Answer

20. (add 1, 1, 2, 2, 3, 3, 4, 4).

44. You have 59 cubic blocks. What is the minimum number that needs to be taken away in order to construct a solid cube with none left over?

Answer

32.

The next cube number below 64 (4 × 4 × 4) is 27 (3 × 3 × 3).

In order to construct a solid cube, therefore, with none left over, 59 – 27 = 32 blocks need to be taken away.

45. If

5862 is to 714
and 3498 is to 1113
and 9516 is to 156
therefore 8257 is to ?

Answer

157.

$7 + 8 = 15$,

$2 + 5 = 7$

46. In a game of eight players lasting for 70 minutes, six substitutes alternate with each player. This means that all players, including the substitutes, are on the pitch for the same length of time. For how long?

Answer

40 minutes.

$(70 \times 8) \div 14$.

Total time for eight players = $70 \times 8 = 560$ minutes. However, as 14 people are each on the pitch for an equal length of time, they are each on the pitch for 40 minutes ($560 \div 14$)

47. 0, 1, 3, 6, 7, 9, 12, 13, 15, 18, ?, ?, ?

What numbers should replace the question marks?

Answer

19, 21, 24.

The sequence progresses +1, +2, +3 repeated.

Chapter 2

How-to-Solve Math Puzzles

1. TRAVELLING TOES

A school bus travels from school to Green Street. There are four children in the bus. Each child has four backpacks with him. There are four dogs sitting in each backpack. Every dog has four puppies with her. All these dogs have four legs, with four toes at each leg. Can you count the total number of toes in the bus?

Answer

The puppies:

4 children × 4 backpacks × 4 dogs × 4 puppies × 4 legs × 4 toes = 4096

Plus the dogs:

4 children × 4 backpacks × 4 dogs × 4 legs × 4 toes = 1024

Plus the children:

4 children × 2 legs × 5 toes = 40

Plus the driver of the school bus:

2 legs × 5 toes = 10

4096 + 1024 + 40 + 10 = 5170

So, the total number of toes is 5170.

2. TRAIN TROUBLE

Shalok walks over a railway-bridge. At the moment that he is just ten metres away from the middle of the bridge, he hears a train coming from behind. At that moment, the train, which travels at a speed of 90 km/h, is exactly as far away from the bridge as the bridge measures in length. Without hesitation, Shalok rushes straight towards the train to get off the bridge. In this way, he misses the train by just four metres. If Shalok had rushed exactly as fast in the other direction, the train would have hit him eight metres before the end of the bridge. What is the length of the railway-bridge?

Answer

Let the length of the bridge be x metres.

Running towards the train, Shalok covers $0.5x - 10$ metres in the time that the train travels $x - 4$ metres.

Running away from the train, Shalok covers $0.5x + 2$ metres in the time that the train travels $2x - 8$ meters.

Because their speeds are constant, the equation formed is as follows:

$(0.5x - 10) / (x-4) = (0.5x + 2) / (2x - 8)$ which can be rewritten as:

$2 \times 0.5x - 24x + 88 = 0$

By solving the quadratic equation we find that $x = 44$. Hence, the length of railway-bridge is 44 metres.

3. KARAN'S HOUSE NUMBER

Karan lives in a street with house-numbers 8 up to 100. Sahil wants to know at which number Karan lives.

He asks him: "Is your number larger than 50?"

Karan Answers, but lies.

Upon this, Sahil asks: "Is your number a multiple of 4?"

Karan Answers, but lies again.

Then Sahil asks: "Is your number a square?"

Karan Answers truthfully.

Upon this, Sahil says: "I know your number if you tell me whether the first digit is a 3."

Karan Answers, but now we do not know whether he lies or speaks the truth.

Thereupon, Sahil says at which number she thinks Karan lives, but (of course) he is wrong.

What is Karan's real house-number?

Answer

Note that Sahil does not know that Karan sometimes lies. Sahil reasons as if Karan speaks the truth. Because Sahil says after his third question, that he knows his number if he tells him whether the first digit is a 3, we can conclude that after his first three questions, Sahil still needs to choose between two numbers, one of which starts with a 3. A number that starts with a 3 must, in this case, be smaller than 50, so Karan's (lied)

Answer to Sahil's first question was "No". Now there are four possibilities:

	number is a square	number is not a square
number is a multiple of 4	16, 36	8, 12, 20, and more
number is not a multiple of 4	9, 25, 49	10, 11, 13, and more

Only the combination "number is a multiple of 4" and "number is a square" results in two numbers, of which one starts with a 3. Karan's (lied) Answer to Sahil's second question therefore was "Yes", and Karan's (true) Answer to Sahil's third question was "Yes" too.

In reality, Karan's number is larger than 50, not a multiple of 4, and a square. Of the squares larger than 50 and at most 100 (these are 64, 81, and 100), this only holds for 81.

Hence, Karan's real house-number is 81.

4. WALKING THE BLANK

The number below is intended to be a 28-digit number, but ten of the digits have been blank. These blanks are to be filled with the digits 0, 1, 2, 3, 4, 5, 6, 7, 8 and 9 – which for the record, can be done in 10! = 3,556,800 different ways. What is the probability that the resulting 28-digit number will be divisible by 396?

5_383_8_2_936_5_8_203_9_3_76

Answer

The probability is one. No matter how you fill in the digits, the resulting 28-digit number will be divisible by 396.

To see why this works, note that $396 = 4 \times 9 \times 11$. At this point we simply need to know the proper divisibility tests for 4, 9and 11, which are stated below.

4: The last two digits form a number that is divisible by 4.

9: The sum of the digits must be divisible by 9.

11: The sum of the digits in odd positions minus the sum of the digits in even positions must be divisible by 11.

Note that the number ends in 76, which is divisible by 4 and therefore, so is the entire number. And the sum of the existing 18 digits equals to 90, which is divisible by 9, as is the sum of 0 through 9, so the whole number is divisible by 9.

It only remains to check for divisibility by 11. The sum of the "odd" digits is 5+ 3 + 3 + 8 + 2 + 9 + 6 + 5 + 8 + 2 +3 +9 +3 +7 = 72, while the sum of the "even" digits is 8 + 3 + 0 + 6 + (sum of 1 through 9) 45 = 62.

The difference of 73 and 62 is 11, which is divisible by 11.

The entire number must always be divisible by 396, no matter where the digits 0 through 9 are placed.

5. TOUGH AGE PUZZLE

My grandson is about as many days as my son in weeks, and my grandson is as many months as I am in years. My grandson, my son and I together are 120 years. Can you tell me my age in years?

Answer

Let my age be m years. If n is my son's age in years, then my son is $52n$ weeks old. If o is my grandson's age in years, my grandson is $365o$ days old. Thus,

$365o = 52n.$

Since my grandson is $12o$ months old,

$12o = m.$

Since my grandson, my son and I together are 120 years,

$o + n + m = 120.$

The above system of 3 equations in 3 unknowns (m, n and o) can be solved as follows:

$\Rightarrow \; m / 12 + 365\, m / (52 \times 12) + m = 120$

$\Rightarrow \; 52\, m + 365\, m + 624\, m = 624 \times 120$

$\Rightarrow \; m = 624 \times 120 / 1041 = 72$

So, my age is 72 years.

6. LETTER PUZZLE

A A D F J J J M M N ?

What comes next in the series?

Answer

The series contains the first letter of each month in alphabetical order (April, August, December, February, January, July, June, March, May, and November). O and S represent the remaining 2 months, October and September.

7. PYRAMID

Given

1 = 5

2 = 25

3 = 325

4 = 4325

What does 5 equals to?

Answer

The Answer is derived by prepending (putting in front of) the value on the left side of the equation with the previous Answer. So for 3 you take 25, the previous Answer, and put 3 in front, giving you 325. Then you put 4 in front to get 4325 and lastly, put 5 in front to get 54325.

8. WHAT'S THE CODE?

Last winter I found myself locked out of my house because I couldn't remember the 5-digit code to open the garage door. I used the following facts to get inside.

1. The second and third digits add up to 9.
2. The first digit is equal to the second digit cubed.
3. The sum of the third and fifth digits is the smallest number with exactly five divisors.
4. The fourth digit is equal to 6 times the second-to-last digit.
5. None of the digits repeat.

What was the code?

Answer

If we label each digit *a*, *b*, *c*, *d* and *e*, we get the following equations:

1. $b + c = 9$
2. $a = (b)^3$
3. $c + e = 16$ (since 16 is the smallest number with five divisors - 1, 2, 4, 8 and 16).
4. $d = 6 \times d$ (The fourth and second-to-last digits are the same number, meaning *d* must be zero to satisfy the equation)
5. *b* must be zero, one or two (*b* can't be 3 because that makes $a = 27$, which isn't a single digit). Zero and one result in duplicate digits (00907 and 11808 respectively)

So, the only remaining value for *b* is 2, giving us 82709.

9. ENTERING THE DOOR

You watch a group of words going to a party. A word either enters through one of two doors or is turned away by the guards. 'HIM' goes through door number one and 'BUG' goes through door number two. 'HER' is turned away. 'MINT' and 'WEAVE' go in through door one, 'DOOR' and 'CORD' take door two and 'THIS' and 'THAT' aren't allowed in. What determines whether a word can enter and which door they must use?

Answer

Door number one is for words composed entirely of capital letters written using only straight lines, such as A, E, F, H, and I. The entire set of letters allowed through door number one is AEFHIKLMNTVWXYZ. Door number two, as might be expected, is for words with capital letters that have a curve, including BCDGJOPQRSU. Any words composed of both straight and curved letters (or lowercase letters) are not allowed in. The word 'THAT' would have been sent through door number one, if the letters had been capitalised.

10. FILL IN NUMBERS

Fill in numbers such that the sum of each 3-character is 22 (row & column) while the sum of all five numbers is 30.

	a	
b	c	d
	e	

Answer

The problem works out to a set of three equations:

$b + c + d = 22$

$a + c + e = 22$

$a + b + c + d + e = 30$

Solving for c = 14, leaving d = 8 - b and e = 8 - a. In other words, c must be 14, but the other two numbers just have to add up to 8. The requirement that they be unique rules out 4 + 4, so you're left to choose from the following combinations for b + d and a + e:

0 + 8

1 + 7

2 + 6

3 + 5

11. STICK FIGURES

Observe the given equation, which is false:

1 1 = 1 1 3 3 5 5

1. Can you insert four line segments into this equation to make it correct?
2. Same question, but now you are given only three line segments to work with. To give a break, the equation doesn't have to be exact, but it does have to be accurate to six decimal places.

Answer

1. 1 1 = – 1 1 – 3 3 + 5 5
2. 1 1 = 1 1 3 3 5 5

Yes, that's ð on the left hand side of the second equation. Remarkably, 355/113 equals 3.14159292035…, whose first six decimal places match those of ð (3.1415926535…).

12. CONNECTING DOTS

Nine dots are arranged in a three by three square. Connect each of the nine dots using only four straight lines and without lifting your pen from the paper.

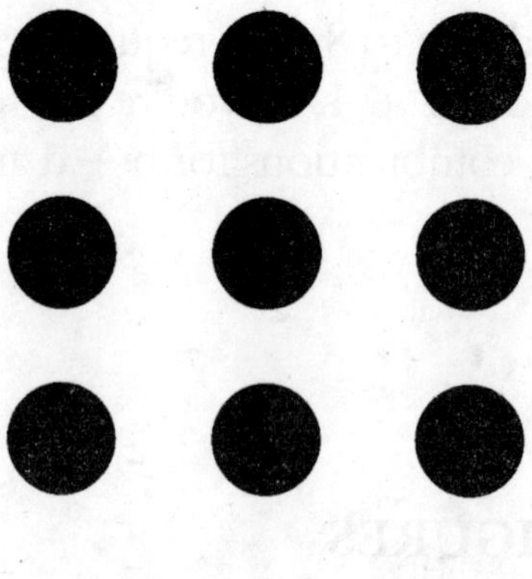

Answer

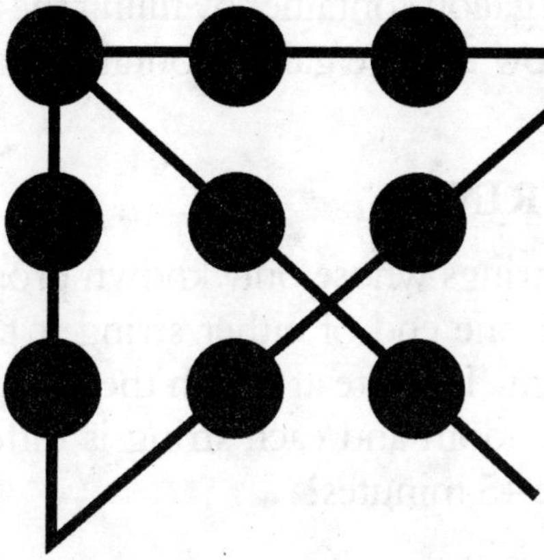

13. DUCKY DUCK

There are two ducks in front of a duck, two ducks behind a duck and a duck in the middle. How many ducks are there in total?

Answer

They're all three ducks in a row. Ducks 1 and 2 are in front of duck 3, ducks 2 and 3 are behind duck 1 and duck 2 is in the middle.

14. FOUR GALLONS

You have a three gallon and a five gallon measuring device. You wish to measure out four gallons.

Answer

Fill the five gallon container. Pour all but two gallons into the three gallon container. Empty the three gallon container. Put the two remaining gallons from the five gallon container into the three gallon container. Fill the five gallon container one more time. Pour one gallon from the five gallon container by filling the three gallon container. Now the five gallon container contains four gallons.

15. TWO STRINGS

You have two strings whose only known property is that when you light one end of either string it takes exactly one hour to burn. The rate at which the strings will burn is completely random and each string is different. How do you measure 45 minutes?

Answer

Light both the ends of the first string and one end of the second string. 30 minutes will have passed when the first string is fully burned, which means 30 minutes have burned off the second string. Light the end of the second string and when it is fully burned, 45 minutes will have passed.

16. THE CUBES

A businessman has two cubes on his office desk. Every day he arranges both cubes so that the front faces show the current day of the month. What numbers are on the faces of the cubes to allow this?

Note: You can't represent the day "7" with a single cube with a side that says 7 on it. You have to use both cubes all the time. So the 7th day would be "07".

Answer

Cube 1 has the following numbers: 0, 1, 2, 3, 4, 5

Cube 2 has the following numbers: 0, 1, 2, 6, 7, 8

The 6 doubles as a 9 when turned the other way around. There is no day 00, but you still need the 0 on both cubes in order to make all the numbers between 01 and 09.

Alternate solutions are also possible as follows:

Cube One: 1, 2, 4, 0, 5, 6

Cube Two: 3, 1, 2, 7, 8, 0

17. TURNING ON A DIME

Suppose that the diameter of a dime was one-fourth the diameter of a silver coin (though in real life the dime is bigger than the silver coin).

If you placed the dime at the top of the silver coin and rotated it around the coin's circumference without slippage – returning to the original position – how many complete revolutions would the dime undergo?

Answer

The solution to the question is five. The distance the coin rolls is equal to the distance travelled by its center. The center travels in a path that forms a circle of radius 5. The center travels 10π units, so the smaller coin rolls through 10π units.

Since, this smaller coin has a circumference of 2π:

Units, it makes 5 full revolutions.

In general, if the circumference of the larger circle is N times that of the smaller circle, the number of complete revolutions equals N + 1.

18. THE POT OF BEANS

A pot contains 75 white beans and 150 black ones. Next to the pot is a large pile of black beans. A somewhat demented cook removes the beans from the pot, one at a time, according to the following strange rule: He removes two beans from the pot at random. If at least one of the beans is black, he places it on the bean-pile and drops the other bean, no matter what color, back in the pot. If both beans are white, on the other hand, he discards both of them and removes one black bean from the pile and drops it in the pot.

At each turn of this procedure, the pot has one less bean in it. Eventually, just one bean is left in the pot. What colour is it?

Answer

The cook only ever removes the two white beans at a time, and there are an odd number of them i.e.(75). When the cook gets to the last white bean, and picks it up along with a black bean, the white one always goes back into the pot. So, the bean left in the pot will be white in colour.

19. FINDING INTEGERS

Can you find five distinct integers A, B, C, D and E, each less than 10, so that the equation below is true?

$A^2 + B^2 + C^2 + D^2 = E^2$

Answer

A = 2, B = 4, C = 5, D = 6 and E = 9 as 4 + 16 + 25 + 36 = 81

20. THE FROG

A frog is at the bottom of a 30 metre well. Each day he summons enough energy for one 3 metre leap up the well. Exhausted, he then hangs there for the rest of the day. At night, while he is asleep, he slips 2 metres backwards. How many days does it take him to escape from the well?

Note: Assume after the first leap that his hind legs are exactly three meters up the well. His hind legs must clear the well for him to escape.

Answer

Each day he makes it up another metre, and then on the twenty eighth day he can leap three metres and climb out.

21. THERE IS SOMETHING ABOUT MARY

Mary's mum has four children.

The first child is called April.

The second May.

The third June.

What is the name of the fourth child?

Answer

Mary's mother's fourth child was Mary herself.

22. DICE GAME

A solo dice game is played where, on each turn, a normal pair of dice is rolled. The score is calculated by taking the product, rather than the sum, of the two numbers shown on the dice.

On a particular game, the score for the second roll is five more than the score for the first; the score for the third roll is six less than that of the second; the score for the fourth roll is eleven more than that of the third; and the score for the fifth roll is eight less than that of the fourth. What was the score for each of these five throws?

Answer

10 is the score for the first roll.

15 is the score for the second roll.

9 is the score for the third roll.

20 is the score for the fourth roll.

12 is the score for the fifth roll.

23. FIVE SQUARES TO TWO

Using just two straight cuts, divide the figure given into three pieces and resemble those pieces to form a rectangle twice as long as it is wide.

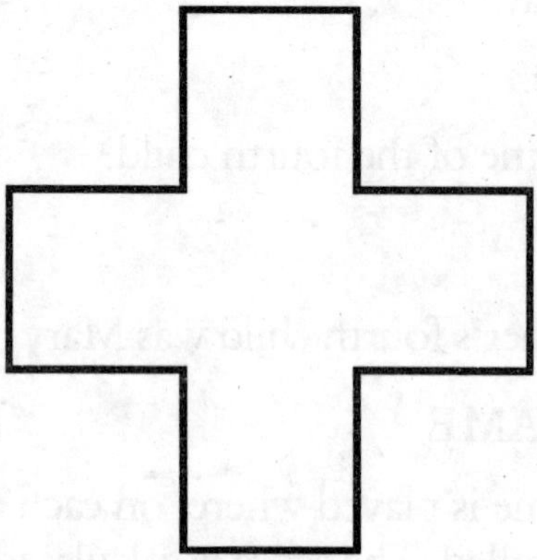

Answer

24. HORSESHOE PATTERN

Arrange the numbers 1 through 15 in a horseshoe pattern in such a way that any two consecutive numbers add up to a perfect square.

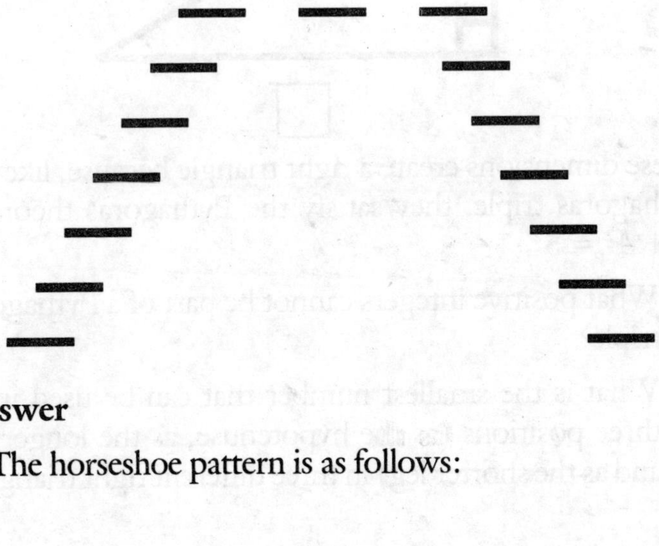

Answer

The horseshoe pattern is as follows:

13 12 4

3 5

6 11

10 14

15 2

1 7

8 9

25. PYTHAGORAS TRIPLE

A Pythagoras triple is a set of three positive integers that can form the sides of a right triangle. The best known example is the famous 3-4-5 right triangle given below:

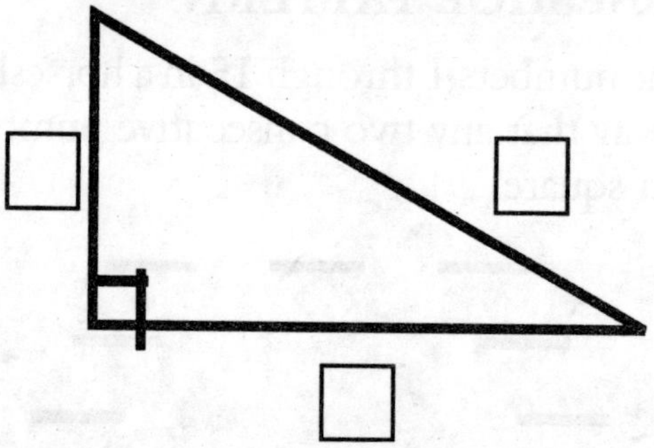

These dimensions create a right triangle because, like any Pythagoras triple, they satisfy the Pythagoras theorem: $3^2 + 4^2 = 5^2$

1. What positive integers cannot be part of a Pythagoras triple?
2. What is the smallest number that can be used in all three positions (as the hypotenuse, as the longer leg and as the shorter leg) in three different right triangles?

Answer

1. The only numbers that cannot be part of a Pythagoras triple are 1 and 2 as there are no two perfect squares that differ by either 1 or 4.

   Let us see why any other number n must be part of a triple.

   First suppose that n is an odd number.

   Then n can be represented as the shorter leg as in the diagram given below:

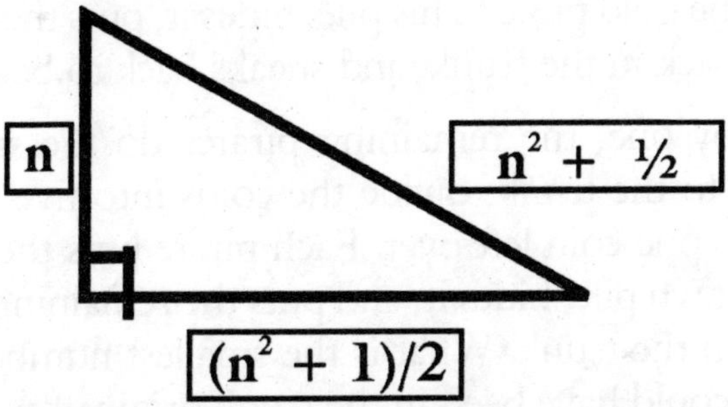

   Note that the longer leg is always one less than the hypotenuse, as in (3, 4, 5), (5, 12, 13) and (7, 24, 25).

   Now for the case where n is an even number.

   If n = 4, we already have a solution – namely, the (3, 4, 5) right triangle. If n is greater than or equal to 6, let n = m × k, with m as odd number.

   We can form a Pythagoras triple by using m, as above. Then we simply multiply all the sides by k.

2. 15 is the smallest such number. It is the hypotenuse of the 9 -12 -15 triangle (obtained by multiplying the 3 – 4 – 5 triangle by 3 on each side); it is the smaller leg of the 15 – 36 – 39 traiangle (obtained by multiplying the 5 – 12 – 13 traingle by 3); and it is the larger leg of the 8 – 15 -17 right triangle.

26. PIRATES COINS

Five pirates raid the ship of a wealthy bureaucrat and steal his trunk of gold pieces. By the time they get the trunk aboard, dusk has fallen, so they agree to split the gold the next morning.

But the pirates are all very greedy. During the night one of the pirates decides to take some of the gold pieces for himself. He sneaks to the trunk and divides the gold pieces into five equal piles, with one gold piece left over. He puts the gold piece in his pile, hides it, puts the other four piles back in the trunk, and sneaks back to bed.

One by one, the remaining pirates do the same. They sneak to the trunk; divide the coins into five piles, with always one coin left over. Each pirate puts the gold coin in his own pile, hides it, and puts the remaining four piles back in the trunk. What is the smallest number of coins there could have been in the trunk originally?

Answer

The original number of coins must be a number such that you can subtract one and multiply by four fifths and get an integer. These numbers are 6, 11, 16, 21, 26, and so on.

But the pile remaining after the first pirate has taken his gold must also have this property. So, the possibilities for the original number are 16, 36, 56, 76, 96, and so on.

The pile remaining after the second pirate has taken his gold must also have this property. So, the possibilities for the original number are 76, 156, 236, 316, 396, and so on.

The pile remaining after the third pirate has taken his gold must also have this property. So, the possibilities for the original number are 316, 636, 956, 1276, 1596, and so on.

The pile remaining after the fourth pirate has taken his gold must also have this property. The smallest possibility for this is 1276.

This number is the number of gold pieces in the chest the fourth pirate left behind (for the fifth pirate to divide). The fourth pirate hid a quarter of this number, plus one extra, just before the fifth pirate got there. So, the third pirate left behind

1276 + 1276/4 + 1 = 1596 gold pieces

The third pirate hid a quarter of this number, plus one extra, just before the fourth pirate got there. So, the second pirate left behind

1596 + 1596/4 + 1 = 1996 gold pieces

The second pirate hid a quarter of this number, plus one extra, just before the third pirate got there. So, the first pirate left behind

1996 + 1996/4 + 1 = 2496 gold pieces

The first pirate hid a quarter of this number, plus one extra. So the original number of coins must have been

2496 + 2496/4 + 1 = 3121 gold pieces

27. HEAD OR TAILS

If you flip a coin five times, what is the probability that three or more consecutive flips come out the same?

Answer

Let us look first at the probability that three or more

heads will come up. There are eight possibilities as follows:

HHHTT, HHHTH, HHHHT, THHHT, THHHH, HTHHH, TTHHH, HHHHH

Similarly, there are eight possibilities with three or more tails. Putting them together, 16 of the 32 possibilities combinations involve three consecutive flips that are the same, so the probability of this event equals to ½.

28. SAVING THE SWAN

A swan is in the center of a circular lake but he cannot take flight from the water, only on land. On the parameter of the lake there is a hunting dog that desperately wants the swan but cannot swim. So the swan must make it to the land before taking off and must do so before the dog makes it to him. The dog is almost 4 times faster than the swan and always runs to the point around the lake closest to the swan. How can the swan get out of the lake and take flight before the dog gets him?

Answer

The swan can travel ¼ of the way to land then swim in a circular path around the center of the lake (the swan will be moving slightly faster around than the dog in their circles). Once the swan is as far as he can get away from the dog in his circle he can swim the remaining ¾ of the way to shore. The dog must travel the radius of the lake time pi (radius × π) while the swan only has to travel ¾ the radius four times slower (¾ × radius × 4). So, the swan will make it to the shore and fly before the dog reaches it.

29. COIN TOSS

Two men find an old gold coin and want to have a coin toss with it to decide who gets it. The only problem is the coin is heavier on one side so it comes up heads more than tails. What is a fair way for the men to toss the coin and decide who gets the coin?

Answer

They just have to flip it twice. They call the first toss heads or tails, then the next toss they automatically pick the opposite (i.e. if one man calls heads on the first flip, he automatically picks tails on the second and vice versa). If they both win one toss (a tie) out of the two, they just have to repeat until one of them wins both tosses

30. QUEENS ON THE BOARD

Queens can move horizontally, vertically and diagonally any number of spaces as illustrated. One piece 'attacks' another if it moves to the same tile that the other piece is on. How can you arrange eight queens on the board so they cannot attack each other?

Answer

Here are the two solutions:

This is usually solved with guess and check although using logic may be faster. We know that each queen must be in its own row vertically and horizontally. We also know that 4 of the queens must be on white and 4 on black. This is true because with 4 queens on the same color all of the rest of that color is venerable to attack.

31. PENTAGON DIVISION

Can you divide the regular pentagon given below into five equal pentagons? The pentagon divisions can be regular or irregular pentagons.

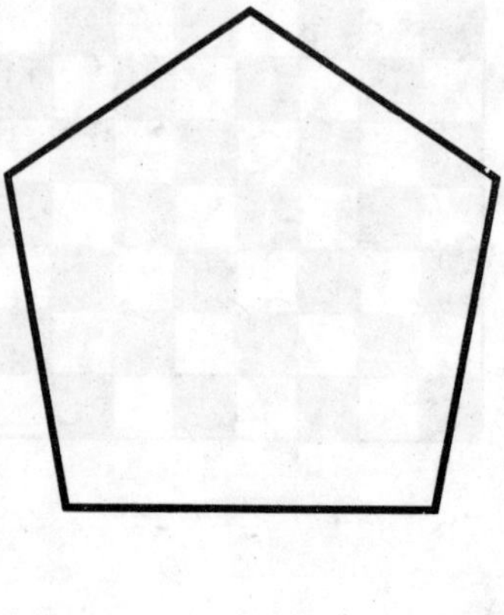

Answer

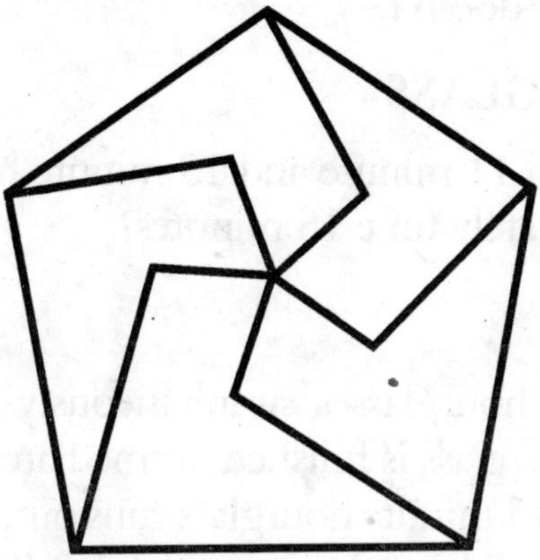

32. NUMBER OF EGGS

Kamal has some chickens that have been laying him plenty of eggs. He wants to give away his eggs to several of his friends, but he wants to give them all the same number of eggs. He figures out that he needs to give 7 of his friends eggs for them to get the same amount, otherwise there is 1 extra egg left. What is the least number of eggs he needs for this to be true?

Answer

He needs 301 eggs.

The number of eggs must be one more than a number that is divisible by 2, 3, 4, 5, and 6 since each of these numbers leave a remainder of 1. For this to be true one less than the number must be divisible by 5, 4, and 3 (6 is 2×3 and 2 is a factor of 4 so, they will automatically be a factor).

$5 \times 4 \times 3 = 60$.

Then you just must find a multiple of 60 such that $60 \times n + 1$ is divisible by 7.

61 / 7, 121 / 7, 181 / 7, 241 / 7 all leave remainders but 301 / 7 doesn't.

33. HOURGLASS

If you have an 11 minute and 13 minute hourglass, how can you accurately time 15 minutes?

Answer

Start both hourglasses simultaneously. When the 11 minute hourglass is finished, immediately flip it again. When the 13 minute hourglass runs out, the 11 minute hourglass will have 9 minutes left, so flip it and it will last another 2 minutes, 13 minutes + 2 minutes = 15 minutes.

34. GOING OFF ON A TANGENT

In the diagram given below, a circle is inscribed in an isosceles trapezoid whose parallel sides have lengths 8 and 18, as indicated. What is the diameter of the circle?

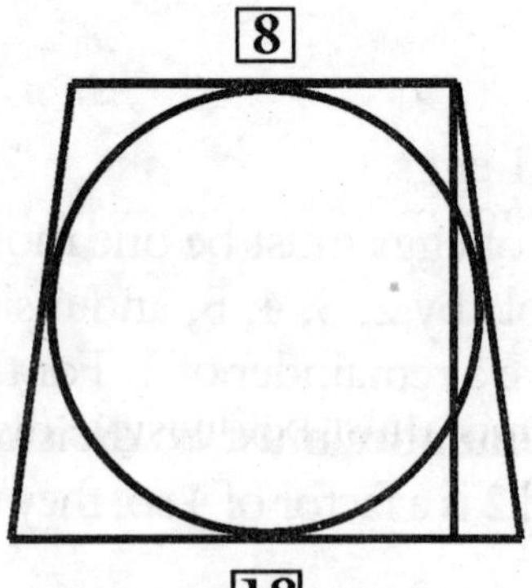

Answer

Any two tangents to a circle must have equal length, so we can segment the lengths of the diagram as follows:

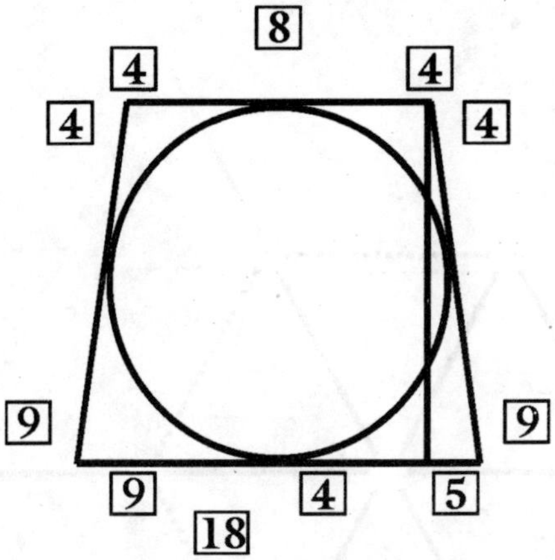

The 5 appears because the length of that segment is the difference of the lengths of two other segments, now known to be 9 and 4, respectively. But the diameter of the circle can now be seen to be a leg of a right triangle with hypotenuse 13 (= 9 + 4) and other leg 5, so by Pythagoras theorem the diameter must be 12.

35. SIX-SIDED STICK SHAPE

The twelve sticks shown on the right form a six-sided shape containing six triangles.

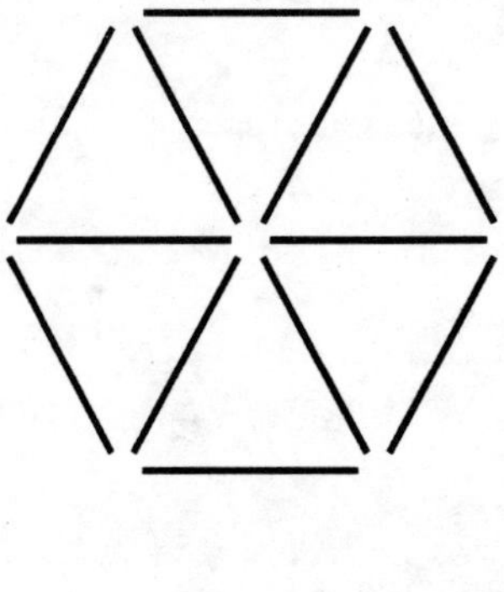

Can you make three triangles by moving four sticks?

Answer

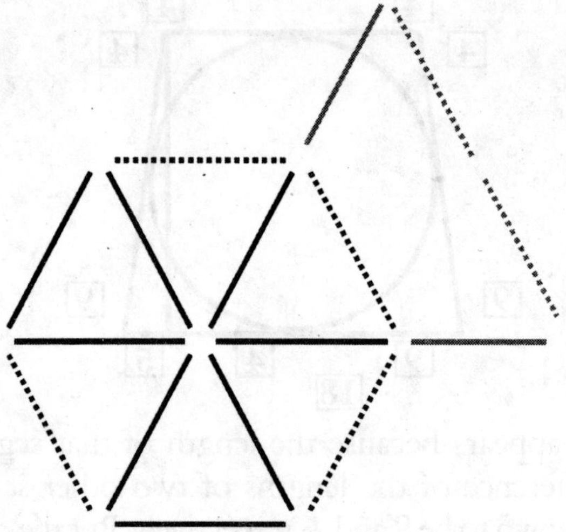

36. COMPLETE THE MULTIPLICATION

In the multiplication shown below, five digits (and the obvious three zeros) are given:

```
 6 . .
 . . . ×
________
 . . .
 . . . . 0
 . 5 . 5 0 0 +
____________
 . . 5 . 4 .
```

What does the complete multiplication look like?

Answer

```
   645
   721 ×
  ------
   645
  12900
 451500 +
 -------
 465045
```

37. ROULETTE TRICK

A well-known roulette trick, to make a profit for sure, is as follows:

You stake continuously at one colour, for example red, double the bet if you lose, and stop as soon as you win. Because you get twice your bet back if you win, and the ball will once fall on red, you know that you will gain your original bet as profit (you must, however, have an infinite amount of money to be able to double your bet every time when necessary). The expected value for your profit is therefore equal to your original bet.

Assume, however, that there is a maximum stake for the roulette, which means that you can only stake n consecutive times with this trick.

Hint : The roulette table has 37 squares: eighteen red, eighteen black and one green. For simplicity, assume that you lose your complete bet if the ball falls on green (French roulette and American roulette have different rules for what happens with your bet in this case).

Answer

Let the original bet be equal to *b*. The probability that you lose a round is 18/37. The probability that you lose *n* consecutive rounds (which means that you must quit with a loss) is (18/37) *n*. Therefore, the probability that you win once is 1-(18/37) *n*. If you lose *n* consecutive rounds, your total bet (and loss) equals (2*n*-1) × b.

If you win once, your profit equals *b*. So, the expected value for your profit is

(18/37)*n* × (-(2*n*-1) × b) + (1-(18/37)*n*) × *b* = (1-(38/37)*n*) × *b*.

Note: This expected value is negative for all values of n greater than 0. You cannot expect a profit, but a loss!

38. THE LONG STRING

Are there ever 1,000,000 consecutive composite numbers?

Answer

The Answer is an emphatic yes. The 1,000,000-term sequence 1,000,001! + 2,1,000,001! + 3, all the way up to 1,000,001! + 1,000,001 consists entirely of composite numbers, because 1,000,001! + K is always evenly divisible by K for any K in this range. More generally, no matter how big the number N is, it is always possible to find N consecutive composite numbers.

39. ANGLED TRIANGLE

We want to find the smallest, right-angled triangle for which holds:

- The lengths of the sides are whole numbers.

- The circumference is the square of a whole number.
- The area is a whole number to the power of three.

Hint: The hypotenuse of the triangle is 240.

Answer

Let the sides of the triangle be a, b, and c, with c = 240 being the hypotenuse.

The triangle has the minimal circumference when a = 1 and b = sqrt (c^2-1^2) (approximately 240). The circumference in that case is approximately 480.

The triangle has the maximal circumference when a and b are equal: $a = b =$ sqrt (½ ×c^2) (approximately 169.7). The circumference in that case is approximately 579.4.

The only two squares of whole numbers that lie in the interval [480.0, 579.4] are 529 and 576.

Now we know that $a + b = 529$ or $a + b = 576$.

In addition, $a^2 + b^2 = c^2$, so $a^2 + b^2 = 57600$.

Suppose that $a + b = 529$. Then $b = 529 - a$, and when we fill that in $a^2 + b^2 = 57600$, we get $a^2 + (529 - a)^2 = 57600$, so $a^2 - 289 \times a + 12960.5 = 0$. This equation has no solutions if a must be integer.

Suppose that $a + b = 576$. Then $b = 576 - a$, and when we fill that in $a^2 + b^2 = 57600$, we get $a^2 + (576 - a)^2 = 57600$, so $a^2 - 336 \times a + 27648 = 0$. This equation has solutions $a = 192$ ($b = 144$) and $a = 144$ ($b = 192$).

Therefore, the sides of the triangle are $a = 144$, $b = 192$, and $c = 240$.

40. TIC-TAC-TOE

Arrange the numbers 1 through 9 on a Tic-Tac-Toe board

such that the numbers in each row, column, and diagonal add up to 15.

Answer

4	3	8
9	5	1
2	7	6

41. THE KING'S GOLD

Long ago, there was a king who had six sons. The king possessed a huge amount of gold, which he hid carefully in a building consisting of a number of rooms. In each room there were a number of chests; this number of chests was equal to the number of rooms in the building. Each chest contained a number of golden coins that equaled the number of chests per room. When the king died, one chest was given to the royal barber. The remainder of the coins had to be divided fairly between his six sons. What is the share of each son?

Answer

A fair division of the coins is indeed possible. Let the number of rooms be N. This means that per room there are N chests with N coins each. In total there are $N \times N \times N = N^3$ coins. One chest with N coins goes to the barber. For the six brothers, $N^3 - N$ coins remain. We can write this as: $N(N^2 - 1)$, or $N(N - 1)(N + 1)$. This last expression is divisible by 6 in all cases, since a number is divisible by 6 when it is both divisible by 3 and even. This is indeed the case here: whatever N may be, the expression $N(N - 1)(N + 1)$ always contains three successive numbers. One of those is always divisible by 3, and at least one of the others is even. This even holds when $N=1$; in that case all the brothers get nothing, which is also a fair division.

42. ELEGANT EQUATION

There is a whole number n for which the following holds: if you put a 4 at the end of n, and multiply the number you get in that way by 4, the result is equal to the number you get if you put a 4 in front of n. In other words, we are looking for the number you can put on the dots in the following equation:

4... = 4 × ...4

Which number must be put on the dots to get a correct equation?

Answer

Make a long division:

4/4.....\......

Calculate the first digit of the result:

4/4.....\1.....

4

—

0

Now we also know that the next digit of the dividend is a 1:

4/41....\1.....

4

—

0

Then we can calculate the next digit of the result:

4/41....\10....

4

—

01

0

—

1

Therefore, the next digit of the dividend is also a 0:

4/410...\10....

4

—

01

0

—

1

This process of calculating the next digit of the result and adding it to the dividend must be repeated until the remainder is 0 *and* the last digit of the result is a 4:

4/410256\102564

4

—

01

0

—

10

8

—

22

20

—

25

24

—

16

16

—

0

Hence, the number we were looking for is 10256.

43. FIND THE SHORTCUT

We know that $5^3 = 125$ and $6^3 = 216$.

With keeping this in mind, suppose you were told that number 148,877 is the cube of some other whole number. What would that other number be?

Answer

The solution to the question is 53. If $5^3 = 125$, then $50^3 = 125{,}000$.

Similarly, $60^3 = 216{,}000$

Note that 148,877 is in between 125,000 and 216,000, so if 148,877 is the cube of some whole number that number must be between 50 and 60. But we know that the cube of a number can end in 7 is if the original number ends in 3.

Therefore, the cube root of 148,877 must be 53.

44. GREEN GRAZING GRASS

A farmer owns a piece of grassland and three animals: a cow, a goat, and a goose. He discovered the following:

- When the cow and the goat graze on the field together, there is no more grass after 45 days.
- When the cow and the goose graze on the field together, there is no more grass after 60 days.
- When the cow grazes on the field alone, there is no more grass after 90 days.
- When the goat and the goose graze on the field together, there is no more grass after 90 days also.

How long the three animals graze together?

Answer

Some assumptions:

The cow, the goat, and the goose eat grass with a constant speed (amount per day): $v1$ for the cow, $v2$ for the goat, $v3$ for the goose.

The grass grows with a constant amount per day (k).

The amount of grass at the beginning is h.

The following is given:

When the cow and the goat graze on the field together, there is no grass left after 45 days. Therefore,

$h - 45 \times (v1 + v2 - k) = 0$, so $v1 + v2 - k = h/45 = 4 \times h/180$

When the cow and the goose graze on the field together, there is no grass left after 60 days. Therefore,

$h - 60 \times (v1 + v3 - k) = 0$, so $v1 + v3 - k = \mathrm{h}/60 = 3 \times h/180$

When the cow grazes on the field alone, there is no grass left after 90 days. Therefore,

$h - 90 \times (v1 - k) = 0$, so $v1 - k = h/90 = 2 \times h/180$

When the goat and the goose graze on the field together, there is also no grass left after 90 days. Therefore,

$h - 90 \times (v2 + v3 - k) = 0$, so $v2 + v3 - k = h/90 = 2 \times h/180$

From this follows:

$v1 = 3 \times h/180$

$v2 = 2 \times h/180$

$v3 = 1 \times h/180$

$k = 1 \times h/180$

Then, holds for the time t that the three animals can graze together:

$h - t \times (v1 + v2 + v3 - k) = 0,$

so $t = h/(v1 + v2 + v3 - k)$

$= h/(3 \times h/180 + 2 \times h/180 + 1 \times h/180 - 1 \times h/180)$

$= 36$

The three animals can graze together for 36 days.

45. TROUBLING TWENTY FOUR

With the digits 1, 4, 5 and 6, you must make 24, using the following rules:

- each digit must be used exactly once
- the allowed operations are addition, subtraction, multiplication, and division (note: exponentiation is therefore not allowed!)
- digits may not be concatenated (so, for example, it is not allowed to use 1 and 4 as 14)
- brackets are allowed

Answer

The two solutions are: 6/(5/4-1) and 4/(1-5/6).

46. CONFUSING CLOCK

The story goes that when the Mahesh did not have so much experience yet with making clocks, a painful mistake was made with a church clock. The clock was officially put into use when it showed 6 o'clock. Soon it was noticed that the hour hand and minute hand had been interchanged and attached to the wrong axes. The result was that the hour hand moved with a speed twelve times higher than the minute hand. When the clock maker arrived, a remarkable thing happened: on the moment he inspected the clock, it showed exactly the right time again. If the clock started at 6 o'clock in the correct position, then what was the first moment that it showed the correct time again?

Answer

Suppose that a second pair of hands turns together with the wrong pair of hands, but then in the correct way. When the wrong pair is in the same position as the correct pair, this means that the time is shown in the right way. First, look at the minute hands that are at the twelve. The "wrong" hand turns twelve times slower than the "correct" hand.

Let x be the distance (in minutes) that the "wrong" hand has progressed when the two minute hands are in the same position again. The "correct" hand then has progressed $60 + x$ minutes (one complete round more). So, it then holds that $12x = 60 + x$. This means that $x = 5\frac{5}{11}$ minutes.

For the hour hands that start at six holds the same. The confused clock therefore shows the correct time again after $60 + 5\frac{5}{11}$ minutes, so at $5\frac{5}{11}$ minutes past 7.

47. LEANING LADDER

In the picture given below, you see a ladder with a length of four metres, placed against a wall. The ladder touches the box of one by one metre, which is standing against the wall. At what height does the top of the ladder touch the wall?

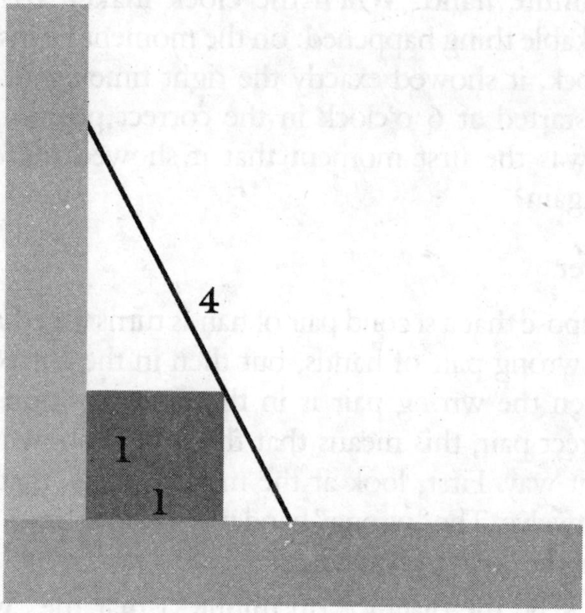

Answer

The picture given below shows the situation.

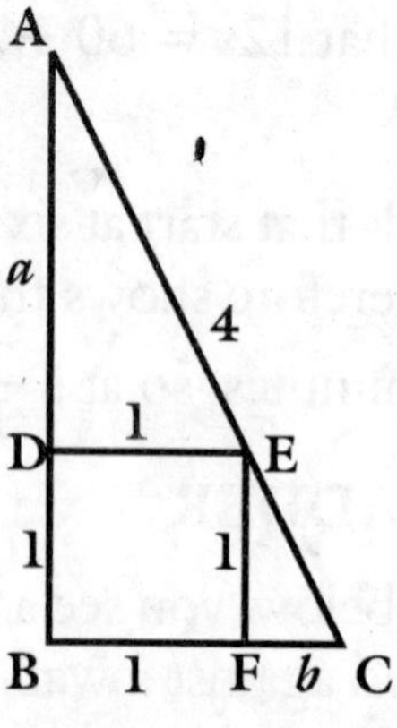

Let a be the length of line segment AD, and let b be the length of line segment CF.

Because of the similarity of the triangles ADE and EFC, the following holds:

$a : 1 = 1 : b$

so $ab = 1$

According to the Pythagorean Theorem, the following holds:

$(AB)^2 + (BC)^2 = (AC)^2$

So $(a + 1)^2 + (1 + b)^2 = 4^2$ which can be rewritten to

$a^2 + 2 + b^2 + 2(a + b) = 16$

Now we use the fact that $ab=1$, so $2 = 2ab$, and we get:

$a^2 + 2ab + b^2 + 2(a + b) = 16$ which can be rewritten to

$(a + b)^2 + 2\times(a + b) - 16 = 0.$

Because we know that $a + b$ is greater than 0, using the quadratic formula we find that $a + b = \text{sqrt}(17) - 1$

Because of the similarity of the triangles ADE and EFC, the following holds:

$a : 1 = 1 : b$

so $b = 1 / a$

Now we know that

$a + 1/a = \text{sqrt}(17) - 1$

so $a^2 + (1 - \text{sqrt}(17)) \times a + 1 = 0.$

We know that a is greater than 0, and using the quadratic formula we find the following two values for ***a***:

½ × (sqrt(17) - 1 + sqrt((1 - sqrt(17))² - 4)) ~
2 . 7 6

½ × (sqrt(17) - 1 - sqrt((1 - sqrt(17))² - 4)) ~
0.36

The ladder touches the wall 1 metre higher, which is at about 3.76 or 1.36 metres. In the figure, we can see that only the **Answer** 3.76 can be correct. So, the ladder touches the wall at about 3.76 metres.

48. THE STAMP COLLECTION

Zeba has a stamp collection consisting of three books. The first book contains 1/5 of the total number of stamps. The second book contains some number of sevenths of the total number of stamps (she doesn't remember how much), and the third book contains 303 stamps. How many stamps are there in the entire collection?

Answer

We know that the number of stamps in the collection is divisible by 35 since the number of stamps can be divided evenly by 5 (a fifth of the stamps are in Book 1) and 7 (some number of sevenths are in Book 2) and because 5 and 7 have no common factor.

Suppose there are $x/7$ of the stamps in the second book. Together, the first and second books contain $1/5 + x/7 = (7 + 5x)/35$ of the collection. The third book therefore contains $(28 - 5x)/35$ of the collection. If there are C stamps in all, $35 \times 303 = C \times (28 - 5x)$. but 35 divides into C, so $(28 - 5x)$, which is a positive integer, must equal one of the factors of 303: 1, 3, 101 or 303. Try each of these cases. The only one that leads to a positive integerless than 7 is $(28 - 5x) = 3$, giving $x = 5$. This yields that 303 stamps amounts to 3/35 of the collection, so the entire collection equals $35 \times 101 = 3535$ stamps.

49. NINETEEN NUMBERS NET

This number net has nineteen circles that have to be filled with the numbers 1 up to (and including) 19. These numbers have to be placed in such a way that all numbers on any horizontal row and any diagonal line add up to the same sum.

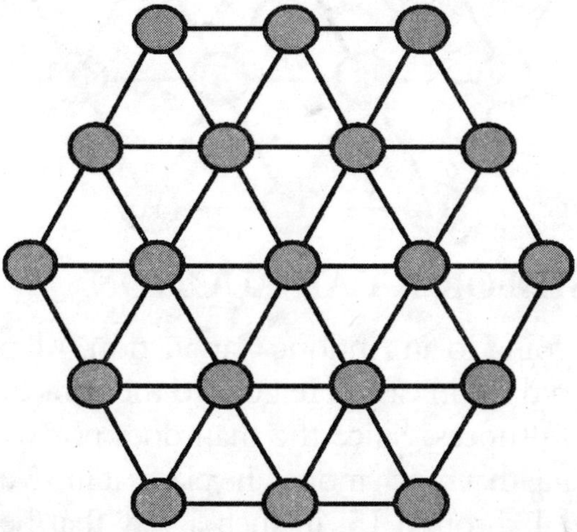

Warning: there are many horizontal and diagonal lines, which have a different number of circles (3, 4, or 5), nevertheless all these sums have to be equal! How should the nineteen numbers be placed in the net?

Answer

As you probably found out, finding the right position for all nineteen numbers in the net is certainly not an easy task. There are 12 possible solutions, but in fact, they are all rotation and mirror symmetrical variants of the same solution. This solution has a sum of 38 on any of its horizontal and diagonal rows, and is shown below:

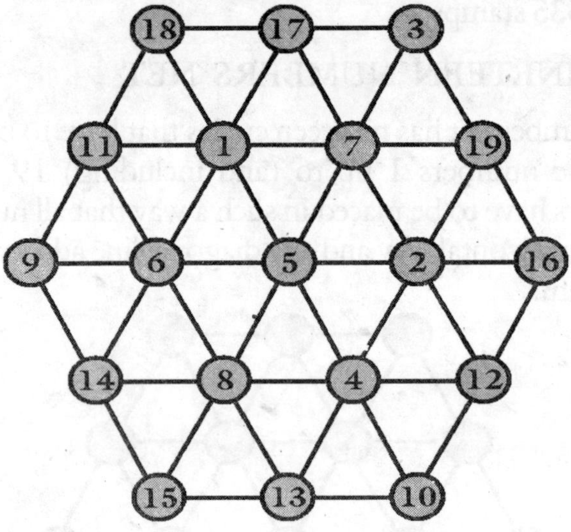

50. CASH FOR A CAR AUCTION

A man is going to an Antique Car auction. All purchases must be paid for in cash. He goes to the bank and draws out 25,000 rupees. Since the man does not want to be seen carrying that much money, he places it in 15 envelopes numbered 1 through 15, in such a way that he can pay any amount up to 25,000 rupees without having to open any envelope. Each envelope contains the least number of bills. At the auction, he makes a successful bid of 8322 rupees for a car. He hands the auctioneer envelopes 2, 8, and 14. After opening the envelopes, the auctioneer finds exactly the right amount. Can you tell how?

Answer

In the envelopes numbered 1 up to 15, the man placed the following amounts of money: 1, 2, 4, 8, 16, 32, 64, 128, 256, 512, 1024, 2048, 4096, 8192, 8617.

The amount of money in each envelope is 2 ^ (number of envelope - 1), except for envelope 15, which contains 8617.

So, the auctioneer finds the right amount i.e. Envelope 2 + 8 + 14 = 2 + 128 + 8192 = 8322

51. SOCK SEARCH

In your bedroom, you have a drawer with 2 red, 4 yellow, 6 purple, 8 brown, 10 white, 12 green, 14 black, 16 blue, 18 grey, and 20 orange socks. It is dark in your bedroom, so you cannot distinguish between the colours of the socks.

How many socks do you need to take out of the drawer to be sure that you have at least three pairs of socks of the same colour?

Answer

In the worst case in which you did not take three pairs of socks of the same colour, you took 2 red socks, 4 yellow socks, and 5 of each of the other colours. That gives a total of 46 socks. Then if you take one more sock, you are sure to have 6 socks of at least one colour. Therefore, you have to take 47 socks from the drawer to be sure that you have at least three pairs of socks of the same colour.

52. FRIDAY THE 13TH

Many people would think Friday the 13th will be an

unlucky day. Is it possible that there is no Friday on 13th through the whole year? How many Fridays at 13th can we have in a year at most? Can you calculate it out?

Answer

We can calculate out how many days there will be for the 13th on each month if we count from the beginning of the year (January 1). Then we divide total days by 7 to get the remainders. We also need to consider the leap year. Through the whole year we had all kinds of remainders, from 0 to 6. The minimum of occurrence for all the unique remainders was 1. It means that we have at least one Friday on 13th. In a regular year, the best chance you can get 3 Fridays on 13th, which are in February, March and December because the remainders of these 3 months are 2. In a leap year, the best chance you also can get 3 Fridays on 13th, which are in January, April and July because the remainders of these 3 months are 6.

53. TURNING CARDS

The following four cards sit on a table:

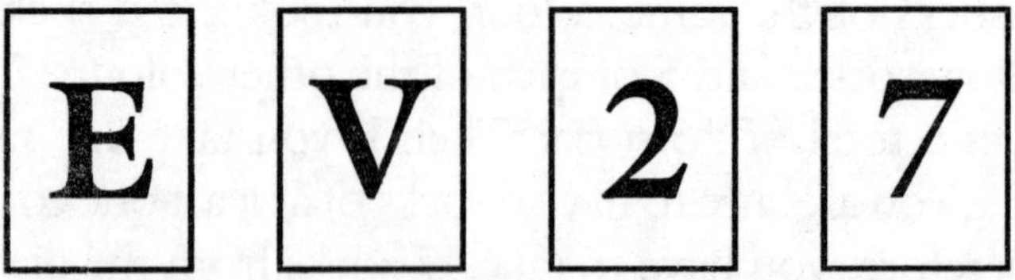

Each card has a digit on one side and a letter on the other side. Which cards should you turn around to test the following statement: "when there is a vowel on one side of a card, then there is an even digit on the other side"?

Answer

You should turn the cards with the "E" and the "7".

The "E" should be turned to verify that there is an even digit on the other side. When there is an odd digit on the other side, the statement is not true.

The "7" should be turned to verify that there is no vowel on the other side. When there is a vowel on the other side, the statement is not true.

The "V" does not need to be turned; it is not a vowel and therefore it does not matter what kind of digit is on the other side.

The "2" also does not need to be turned. Whether there is a vowel or a consonant on the other side, the card always satisfies the statement: after all, it is not stated that only cards with a vowel have an even digit on the other side!

54. NUMBER TRICK

Sonal asked the class to see if they could find the sum of the first 50 odd numbers. As everyone settled down to their addition, Lavesh ran to her and said, 'The sum is 2,500.' Ms. Sonal thought, 'Lucky guess,' and gave him the task of finding the sum of the first 75 odd numbers. Within 20 seconds, Lavesh was back with the correct Answer.

How does Lavesh find the sum so quickly and what is the Answer?

Answer

The following pattern holds: The sum is equal to n x n,

when n is the number of consecutive odd numbers, starting with 1. For example, the sum of the first 3 odd numbers is equal to 3 x 3, or 9; the sum of the first 4 odd numbers is equal to 4 x 4, or 16; the sum of the first 5 odd numbers is equal to 5 x 5, or 25; and so on... The Sum of first 75 odd no's is 75x75 or 5625.

55. THE ROUND TABLE

Yesterday evening, Helen and her husband invited their neighbours (two couples) for a dinner at home. The six of them sat at a round table. Heena tells you the following:

"Vicky sat on the left of the woman who sat on the left of the man who sat on the left of Anita.

Jai sat on the left of the man who sat on the left of the woman who sat on the left of the man who sat on the left of the woman who sat on the left of my husband.

Kapil sat on the left of the woman who sat on the left of Rajeev. I did not sit beside my husband." What is the name of Heena's husband?

Answer

From the second statement, we know that the six people sat at the table in the following way (clockwise and starting with Heena's husband):

Heena's husband, woman, man, woman, man, Jai

Because Heena did not sit beside her husband, the situation must be as follows:

Heena's husband, woman, man, Heena, man, Jai

The remaining woman must be Anita, and combining this with the first statement, we arrive at the following situation:

Heena's husband, Anita, man, Heena, Vicky, Jai

Because of the third statement, Kapil and Rajeev can be placed in only one way, and we now know the complete order:

Heena's husband Rajeev, Anita, Kapil, Heena, Vicky, Jai

Hence, the name of Heena's husband is Rajeev.

56. SQUARE CIRCLES

Given are the following three equations:

■ ● = ▲

■ = ● ◆

▲ ▲ = ◆ ◆ ◆

How many circles is a square, if you take the ratios in the three given equations; in other words: how many circles should be on the dots below?

■ = ...

Answer

■ ■ ■ = ● ◆ ● ◆ ● ◆

= { according to equation 2 }

= sorting them, we get—

● ● ● ◆ ◆ ◆

= By Using equation 3, we get—

● ● ● ▲ ▲

= ● ● ● ■ ● ■ ● { according equation 1 }

Again sorting them, we get—

■■●●●●●

Consequently, ■■■ = ■■●●●●●

from which follows that ■ = ●●●●●

Hence, the Answer is five circles.

57. GOLD BAR FEWER CUT

A worker is to perform work for you for seven straight days. In return for his work, you will pay him 1/7th of a bar of gold per day. The worker requires a daily payment of 1/7th of the bar of gold. What and where are the fewest number of cuts to the bar of gold that will allow you to pay him 1/7th each day?

Answer

Day One: You make your first cut at the $1/7^{th}$ mark and give that to the worker.

Day Two: You cut $2/7^{th}$ and pay that to the worker and receive the original $1/7^{th}$ in change.

Day three: You give the worker the $1/7^{th}$ you received as change on the previous day.

Day four: You give the worker $4/7^{th}$ and he returns his $1/7^{th}$ cut and his $2/7^{th}$ cut as change.

Day Five: You give the worker back the $1/7^{th}$ cut of gold.

Day Six: You give the worker the $2/7^{th}$ cut and receive the $1/7^{th}$ cut back in change.

Day Seven: You pay the worker his final $1/7^{th}$.

58. FINDING SQUARES

How many squares are present in the picture of the board?

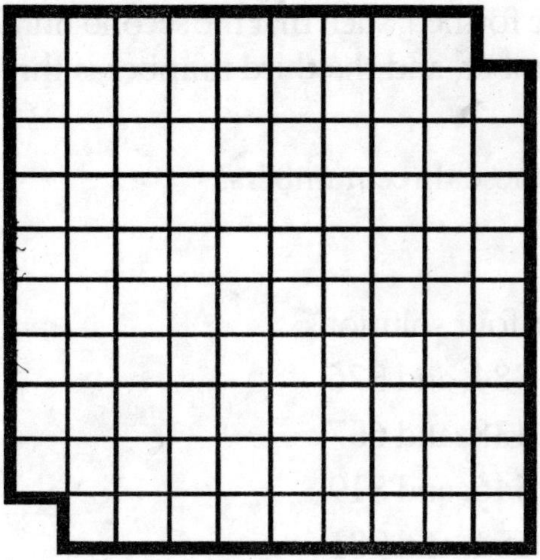

Answer

Here is an overview of the number of squares of each size, as they are present in the figure:

9×9: $2^2 - 2 = 2$

8×8: $3^2 - 2 = 7$

7×7: $4^2 - 2 = 14$

6×6: $5^2 - 2 = 23$

5×5: $6^2 - 2 = 34$

4×4: $7^2 - 2 = 47$

3×3: $8^2 - 2 = 62$

2×2: $9^2 - 2 = 79$

1×1: $10^2 - 2 = 98$

\+ ——

<u>366</u>

Hence, there are 366 squares.

59. ONE, TWO, THREE

Using the digits 1 up to 9, three numbers (of three digits each) can be formed, such that the second number is twice the first number, and the third number is three times the first number.

Which are these three numbers?

Answer

There are four solutions:

- 192, 384, and 576
- 219, 438, and 657
- 273, 546, and 819
- 327, 654, and 981

60. SPLITING SHAPES

The shape shown below must be partitioned into four identical pieces.

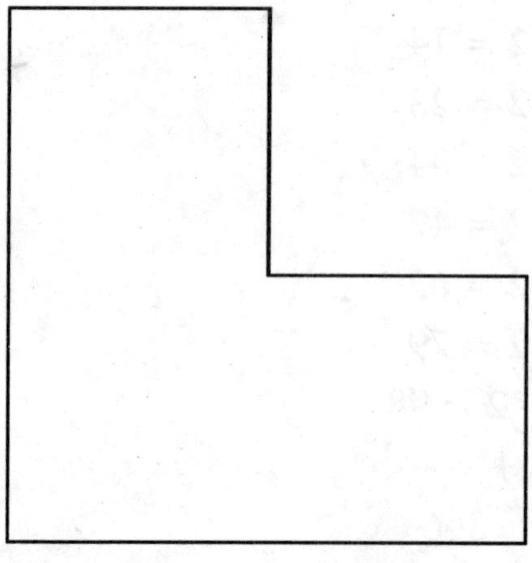

How can this be done?

Answer

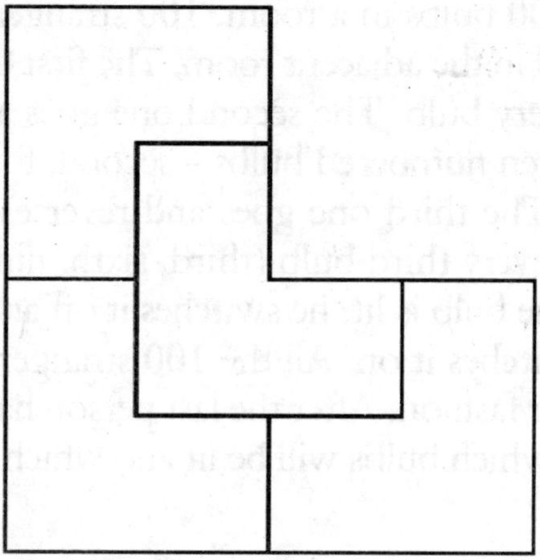

61. SALMAN'S AGE

Salman's youth lasted one sixth of his life. He grew a beard after one twelfth more. After one seventh more of his life, he married. 5 years later, he and his wife had a son. The son lived exactly one half as long as his father and Salman died four years after his son. How many years did Salman live?

Answer

Let Salman's age at the time of his death be x years.

According to the question following equation is formed:

$x/6 + x/12 + x/7 + 5 + x/2 + 4 = x$

Solving the equation we get x as 84.

Hence, Salman lived exactly 84 years.

62. LIGHTINING BULBS

There are 100 bulbs in a room. 100 strangers have been accumulated in the adjacent room. The first one goes and lights up every bulb. The second one goes and switches off all the even numbered bulbs – second, fourth, sixth... and so on. The third one goes and reverses the current position of every third bulb (third, sixth, ninth… and so on.) i.e. if the bulb is lit, he switches it off and if the bulb is off, he switches it on. All the 100 strangers progresses in the similar fashion. After the last person has done what he wanted, which bulbs will be lit and which ones will be switched off?

Answer

Ponder over the bulb number 56, people will visit it for every divisor it has. So, 56 has 1 & 56, 2 & 28, 4 & 14, 7 & 8. So, on pass 1, the 1st person will light the bulb; pass 2, 2nd one will switch it off; pass 4, light it; pass 7, switch it off; pass 8, light it; pass 14, switch it off; pass 28, light it; pass 56, switch it off.

For each pair of divisors the bulb will just end up back in its preliminary state. But there are cases in which the pair of divisor has similar number for example bulb number 16. 16 has the divisors 1 & 16, 2 & 8, 4 & 4. But 4 is recurring because 16 is a perfect square, so you will only visit bulb number 16, on pass 1, 2, 4, 8 and 16… leaving it lit at the end. So, only perfect square bulbs will be lit at the end.

63. COUNTING HAIR

There is a town named Springville. In that town you can

find extremely hilarious facts. None of the resident has 456,789 hairs. The number of residents are more than the number of hair in any one of the. Can you find out the largest number of residents in such a scenario?

Answer

Let us begin with two residents. The number of hairs on their head can be zero and one. If we extrapolate the fact, we will come to know that the number of hairs with 'n' number of residents will always range from zero to (n-1). If we go above 456,789 to 456,790 with none of the residents having 456,789 hairs, the number of hairs on them will be different and one of them must have more than 456,789 which will clearly violate the fact that the numbers of residents are more than hairs on an individual.

Hence, the Answer is 456,789.

64. ARRAY OF NUMBERS

We have arranged an array of numbers below. What you have to do is use any kind of mathematical symbol you know excluding any symbol that contains a number like cube root. You can use any amount of symbols but you have to come up with a valid equation for all of them.

0 0 0 = 6

1 1 1 = 6

2 + 2 + 2 = 6

3 3 3 = 6

4 4 4 = 6

5 5 5 = 6

6 6 6 = 6

7 7 7 = 6

8 8 8 = 6

9 9 9 = 6

Answer

$(0! + 0! + 0!)! = 6$

$(1 + 1 + 1)! = 6$

$2 + 2 + 2 = 6$

$3 \times 3 - 3 = 6$

$\sqrt{4} + \sqrt{4} + \sqrt{4} = 6$

$5 + 5/5 = 6$

$6 + 6 - 6 = 6$

$7 - 7/7 = 6$

$8 - \sqrt{\sqrt{(8 + 8)}} = 6$

$\sqrt{(9 \times 9)} - \sqrt{9} = 6$

65. THE COIN SYSTEM

In the principality of Potluck, the currency is the Fluke which is worth 100 Jammies. The Bank of Potluck has just issued the new One Fluke note, but now want a coin system for the Jammies. The plan is that you should be able to make ANY amount between 1 and 100 Jammies by using just one or two coins. (The two coins can be the same or different.) What is the smallest number of different coins they need for their new system - and what are they?

Answer

They need 16 different types of coin:

1, 3, 4, 9, 11, 16, 20, 25, 30, 34, 39, 41, 46, 47, 49, 50

(These coins will make the amount between 1 and 100 as shown below:

50 + 50 = 100

50 + 49 = 99

49 + 47 = 98 and so on....

We have chosen odd numbers as the even number will make repeated amount)

66. THE MIDNIGHT TRAIN

The four Gabrianni brothers have 17 minutes to catch the midnight train out of town. To reach the station they must all cross a narrow bridge BUT... it's dark, they only have one torch and the bridge can only carry a maximum of TWO people at once.

Any person or couple crossing the bridge MUST have the torch with them. The torch must be carried across, it cannot be thrown. Each brother walks at a different speed and a pair must walk together at the rate of the slower man's pace.

Chainsaw Charlie: 1 minute to cross

Numbers: 2 minutes to cross

Half-Smile: 5 minutes to cross

Weasel: 10 minutes to cross (because they made him carry all the cases.)

For example: If Chainsaw and Weasel cross first it will

take 10 minutes. If Weasel comes back with the torch, that's 20 minutes gone and they missed the train! So can you help them reach the train on time?

Answer

Charlie and Numbers go across (2mins)

Charlie comes back (1min).

Half-Smile and Weasel go across (10 mins)

Numbers comes back (2mins)

Charlie and Numbers go across (2mins).

Total time= 17 mins

67. MAGGIE PRIZE

A Maggie company places prizes in its Maggie packets. There are four different prizes distributed evenly over all the packets that the company produces. On average, how many packets of Maggie would you need to buy before you collected a complete set?

Answer

The Answer is $8\frac{1}{3}$ packets. To see why, note that one of the prizes comes from the first packet we buy. The likelihood of getting new prize from the next packet equals ¾; on average, therefore, we would need to buy 4/3 packets to get a new prize.

Proceeding in this same fashion, the third new prize would require an additional $1/(1/2) = 2$ packets. The fourth would require, on average, an additional 4 packets.

In total, the average number of packets equals $1 + 4/3$

$+ 2 + 4 = 8\frac{1}{3}$ packets. In real life, of course, you can't buy 1/3 of a packet, but that is still average number of boxes you'd have to purchase.

68. THE SEVEN SQUARES

There are 28 spots here. Can you connect the spots with straight lines to make seven squares? You may not use any spot more than once.

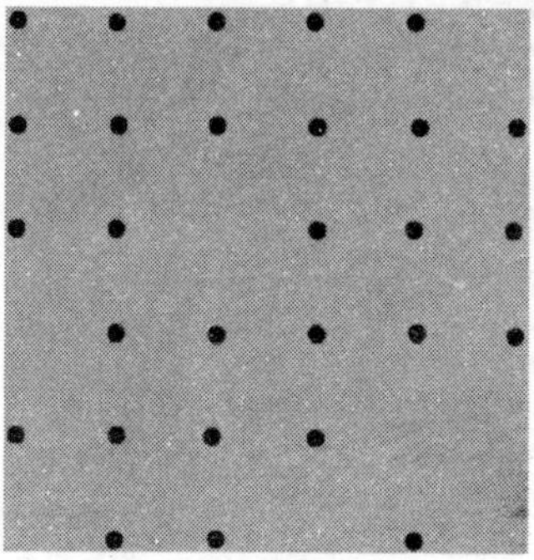

Answer

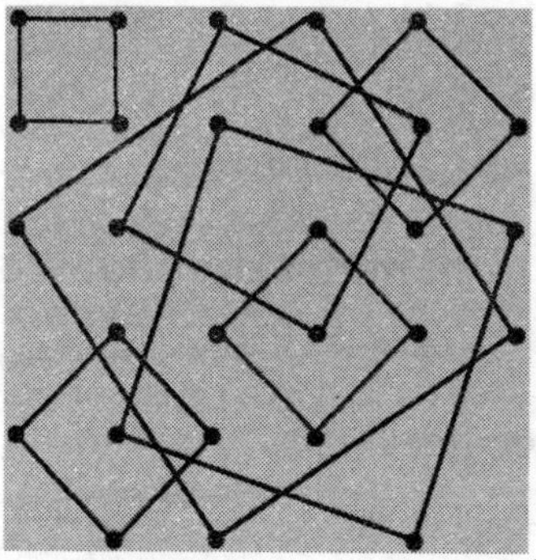

69. AYAAN'S CAGE

There is a cage in which the Ayaan keeps his three tongued monster. The cage door is protected by several different locks. Ayaan and his four slaves were in charge of guarding the door, but his father had given out the keys in such a way that Ayaan could only unlock the door if (any) one of the slaves was with him. The slaves could only unlock the door if at least three of them worked together. So, what was the smallest number of locks needed for the door?

Answer

There were seven locks labelled A B C D E F G.

Ayaan had keys ABCDEF.

The slaves had keys ABCG ADEG BDFG CEFG

70. CAN 2 = 1?

Here's our "proof":

1. Let $x = y$
2. so

   $x^2 = xy$
3. adding x^2 to both sides of the equation we get $x^2 + x^2 = x^2 + xy$
4. simplifying we get

   $2\,x^2 = x^2 + xy$
5. subtract 2xy from both sides and we get

   $2\,x^2 - 2xy = x^2 + xy - 2xy$
6. simplifying we get

   $2\,x^2 - 2xy = x^2 - xy$
7. factoring for $(x^2 - xy)$ we get

   $2\,(x^2 - xy) = 1\,(x^2 - xy)$
8. divide both sides by $(x^2 - xy)$ we get

   $2 = 1$

Since 2 can't equal. 1 there must be something wrong here. What's wrong with our proof?

Answer

The problem with our proof is in step 8:
divide both sides by $(x^2 - xy)$ we get $2 = 1$

In step 1 we said that $x = y$, so $(x^2 - xy) = 0$, and you can't divide by 0.

Our flawed proof is a good example of why dividing by 0 is not allowed. It's not just because your math teacher told you so. It is because dividing by 0 can produce impossible results.

71. STICKING AWAY SQUARES

In the figure given below, can you leave just three squares by taking away three sticks?

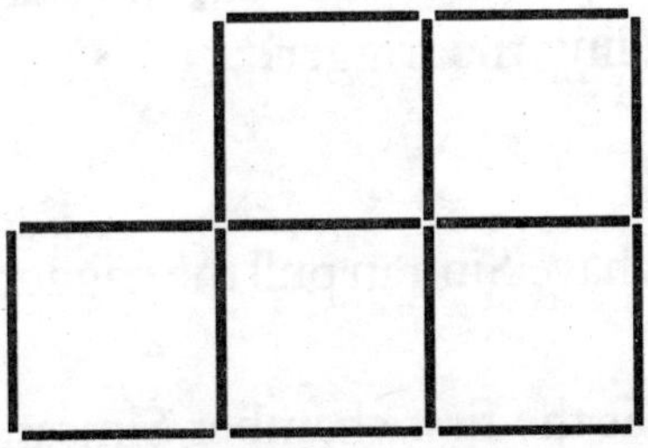

Answer

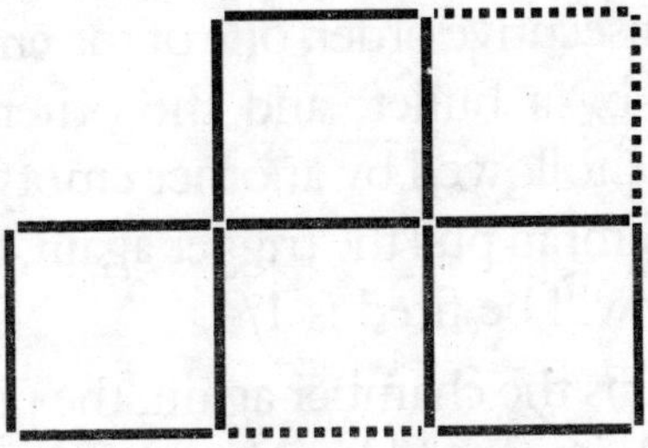

72. RAJAT'S JUSTICE

Rajat has been caught stealing cattle, and is brought into town for justice. The judge is his ex-wife Simran, who wants to show him some sympathy, but the law clearly calls for two shots to be taken at Rajat from close range. To make things a little better for Rajat, Simran tells him she will place two bullets into a six-chambered revolver in successive order. She will spin the chamber, close it, and take one shot. If Rajat is still alive, she will then either take another shot, or spin the chamber again before shooting.

Rajat is a bit incredulous that his own ex-wife would carry out the punishment, and a bit sad that she was always such a rule follower. He steels himself as Simran loads the chambers, spins the revolver, and pulls the trigger. Whew! It was blank. Then Simran asks, 'Do you want me to pull the trigger again, or should I spin the chamber a second time before pulling the trigger?'

Answer

Rajat should have Simran pull the trigger again without spinning.

We know that the first chamber Simran fired was one of the four empty chambers. Since the bullets were placed in consecutive order, one of the empty chambers is followed by a bullet, and the other three empty chambers are followed by another empty chamber. So, if Rajat has Simran pull the trigger again, the probability that a bullet will be fired is 1/4.

If Simran spins the chamber again, the probability that she shoots Rajat would be 2/6, or 1/3, since there are two possible bullets that would be in firing position out of the six possible chambers that would be in position.

73. BEGGAR'S MONEY

On a festival, a beggar earns 71 rupees in form of 20-paisa and 25-paisa. He collects total of 324 coins. Can you tell me how many number of 20-paisa and 25-paisa he got ?

Answer

20 paisa coins : 200

25 paisa coins : 124

Let's see how the **Answer** came.

Say he collect 20-paisa coins be A.

Say he collect 25-paisa coins be B.

Therefore, number of 25-paisa coins = (324 - A)=B.

Therefore,

$0.20 \times A + 0.25\,(324 - A) = 71$

$20A + 25\,(324 - A) = 7100$

$5A = 1000$

$A = 200$

Hence, number of 25-paisa coins = (324 - A) – 124

74. HAVING SAME BIRTHDAY

How many people must be gathered together in a room, before you can be certain that there is a greater than 50/50 chance that at least two of them have the same birthday?

Answer

Only twenty-three people need be in the room, a surprisingly small number. The probability that there will not be two matching birthdays is then, ignoring leap years, is365x364x363x...x343/365 over 23 which is approximately 0.493. this is less than half, and therefore the probability that a pair occurs is greater than 50-50. With as few as fourteen people in the room the chances are better than 50-50 that a pair will have birthdays on the same day or on consecutive days.

75. AEROPLANE AND TANKS OF FUEL

On Bagshot Island, there is an airport. The airport is the homebase of an unlimited number of identical airplanes. Each airplane has a fuel capacity to allow it to fly exactly 1/2 way around the world, along a great circle. The planes have the ability to refuel in flight without loss of speed or spillage of fuel. Though the fuel is unlimited, the island is the only source of fuel.

What is the fewest number of aircraft necessary to get one plane all the way around the world assuming that all of the aircraft must return safely to the airport? How did you get to your Answer?

Notes:

(a) Each airplane must depart and return to the same airport, and that is the only airport they can land and refuel on ground.

(b) Each airplane must have enough fuel to return to airport.

(c) The time and fuel consumption of refueling can be

ignored. (So, we can also assume that one airplane can refuel more than one airplane in air at the same time.)

(d) The amount of fuel airplanes carrying can be zero as long as the other airplane is refueling these airplanes.

What is the fewest number of airplanes and number of tanks of fuel needed to accomplish this work? (we only need airplane to go around the world)

Answer

The fewest number of aircraft is 3.

Imagine 3 aircraft (A, B and C). A is going to fly round the world. All three aircraft start at the same time in the same direction. After 1/6 of the circumference, B passes 1/3 of its fuel to C and returns home, where it is refueled and starts immediately again to follow A and C.

C continues to fly alongside A until they are 1/4 of the distance around the world. At this point C completely fills the tank of A which is now able to fly to a point 3/4 of the way around the world. C has now only 1/3 of its full fuel capacity left, not enough to get back to the home base. But the first 'auxiliary' aircraft reaches it in time in order to refuel it, and both 'auxiliary' aircraft are the able to return safely to the home base.

Now in the same manner as before both B and C fully refuelled fly towards A. Again B refuels C and returns home to be refuelled. C reaches A at the point where it has flown 3/4 around the world. All 3 aircraft can safely return to the home base, if the refuelling process is applied analogously as for the first phase of the flight.

76. REPLACE THE QUESTION MARK

Replace the question mark with the correct number.

1	3	6	9
6	8	7	3
2	4	?	6
6	4	7	2

Answer

6×3 equals 18

7×9 equals 63

6×4 equals 24

7×6 equals 42,

So, ? will be replaced by 4.

77. REMAINING PRODUCT

If we have four positive numbers a, b, c and d, there are six ways to multiply the pairs i.e. $a \times b$, $a \times c$, $a \times d$,

$b \times c$, $b \times d$ and $c \times d$. If we tell you the result of five of them without telling you which one is the product of which pair as 2, 3, 4, 5 and 6. What is the remaining product?

Answer

On multiplying all the 6 pairs we get $(a \times b \times c \times d)^3$.

Assuming the missing product is $a \times b$, then the 5 given products, we see:

$(a \times c) \times (b \times d) = (a \times d) \times (b \times c)$

If we look at the numbers clearly 5 is out of the equation because there is nothing to balance it so the only remaining match is $2 \times 6 = 3 \times 4$, so $a \times b \times c \times d = 2 \times 6 = 12$

So, the product of all 6 pairs should be 12^3, the product of the given 5 pairs is 720, so the missing one is $12^3/720 = 2.4$

78. NINETEEN

How can you write nineteen in a manner that if we take out one, it becomes twenty?

Answer

XIX, on taking one i.e. I, we get XX i.e. 20

79. A HORSE AND A CAMEL

A horse and a camel were carrying boxes on their backs. The horse started complaining to the camel that his load is too heavy. The camel replied 'Why are you complaining? If you gave me one of your boxes I would have double what you have and if I give you one of my boxes we two would have an even load.' How many boxes do each of the animal (horse & camel) is carrying?

Answer

Horse is carrying 5 boxes and the camel is carrying 7 boxes.

Let's assume that the horse was carrying H boxes and the camel was carrying C boxes.

$C + 1 = 2 * (H-1)$

$C + 1 = 2H - 2$

$C = 2H - 3$

Also,

$C - 1 = H + 1$

$C = H + 2$

$\therefore 2H - 3 = H + 2$

$H = 5$

$\Rightarrow C = 7$

80. A HORSE AND A CAMEL

Nitin, Jatin, and Pavit race each other in a 100 metres race. All of them run at a constant speed throughout the race.

Nitin beats Jatin by 20 metres.

Jatin beats Pavit by 20 metres.

How many metres does Nitin beat Pavit by ?

Answer

We let Nitin's speed be A metre/second. So, it takes

him 100/A seconds to finish the race. At this point, we know that Jatin has run 80 metres (since Nitin beats him by 20 metres).

So, Jatin runs 80 metres in 100/A seconds, meaning that he is running at a speed of (80/(100/A)) metres/second, or (8A/10) metres per second.

So, we then know that it takes Jatin 100/(8A/10) seconds to finish the race, or 125/A seconds. At this point, we know that Pavit has run 80 metres (since Jatin beats him by 20 metres).

So, Pavit runs 80 metres in 125/A seconds, meaning that he is running at a speed of (80/(125/A)) metres/second, or 80A/125 metres per second.

Now, that we know Pavit's speed, we just need to figure out how far he had run when Nitin finished the race. Since Nitin finished in 100/A seconds, we can determine that Pavit had run $(100/A) \times (80A/125) = 8000/125 = 64$ metres when Nitin finished the race.

So, Nitin beat him by $(100 - 64) = 36$ metres

81. MAKING NINE

Can you rearrange these three sticks to make nine?

Answer

82. GUESS THE CARDS

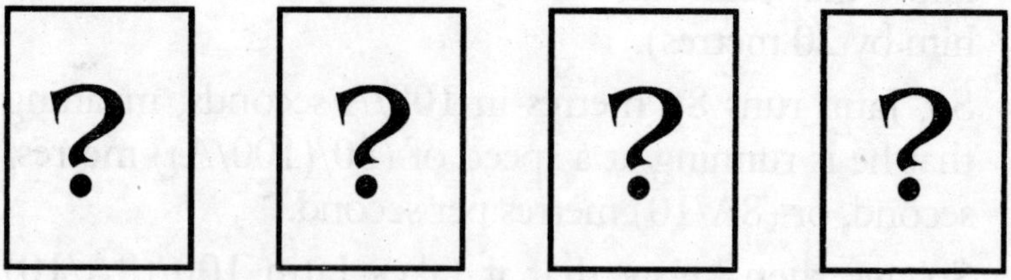

From a pack of 52 cards , i placed 4 cards on the table.

I will give you 4 clues about the cards:

Clue 1: Card on left cannot be greater than card on the right.

Clue 2: Difference between 1st card and 3rd card is 8.

Clue 3: There is no card of ace.

Clue 4: There is no face cards (queen,king,jacks).

Clue 5: Difference between 2nd card and 4th card is 7.

Identify the four cards ?

Answer

2 3 10 10

The 1st card has to be 2 and last card has be 10 as there is no other way difference can be 8.

⇒ 2 ? 10 ?

because of clue 4, we know 4th card is 10

⇒ 2 ? 10 10

because of clue 5, we know 1st card is 3

⇒ 2 3 10 10

83. SMALLEST NUMBER

Which is the smallest number that you can write using all the vowels exactly once?

Answer

FIVE THOUSAND

84. TWO TWINS

Two Twins were standing back to back and suddenly they started running in opposite direction for 4 kilometers and then turn to left and run for another 3 kilometers. What is the distance between the twins when they stop?

Answer

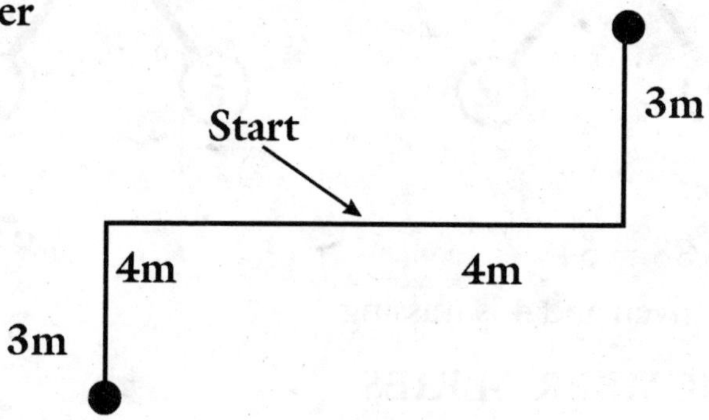

10 kilometres

85. MAGIC NUMBER

Ramanujan discovered 1729 as a magic number. Why 1729 is a magic number ?

Answer

It is a magic number because it can be expressed as the sum of the cubes of two different sets of numbers.

$10^3 + 9^3 = 1729$

and

$12^3 + 1^3 = 1729$

86. CORRECT NUMBER

Replace the question mark with the correct number.

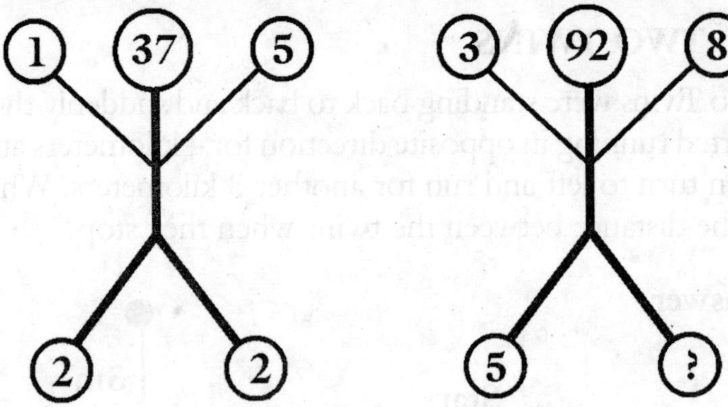

Answer

92 - 38 = 54

5 is given and 4 is missing

87. NUMBER SERIES

Find the next number in the series

21 32 54 87 131 ?

Answer

$21 + 11 \times 1 = 32$

$32 + 11 \times 2 = 54$

$54 + 11 \times 3 = 87$

$87 + 11 \times 4 = 131$

$131 + 11 \times 5 = 186$

Hence, the next number in the series is 186.

88. HANDSHAKE

At a party, everyone shook hands with everybody else. There were 66 handshakes. How many people were at the party?

Answer

12

In general, with n + 1 people, the number of handshakes is the sum of the first n consecutive numbers: 1 + 2 + 3 + ... + n.

Since this sum is n (n + 1)/2, we need to solve the equation n (n + 1)/2 = 66.

This is the quadratic equation $n^2 + n - 132 = 0$. Solving for n, we obtain 11 as the Answer and deduce that there were 12 people at the party.

Since 66 is a relatively small number, you can also solve this problem with a hand calculator. Add 1 + 2 = + 3 = +... etc. until the total is 66. The last number that you entered (11) is n.

89. MATHS EQUATION PUZZLE

How can get total of 120 by using five zeros 0, 0, 0, 0, 0 and any one mathematical operator?

Answer

(0!+0!+0!+0!+0!)!

= (1+1+1+1+1)!

= (4)!

= 5! is 5 × 4 × 3 × 2 × 1 = 120

90. SOLVE IT

2+3=8,

3+7=27,

4+5=32,

5+8=60,

6+7=72,

7+8=??

Solve it?

Answer

$2 + 3 = 2 \times [3 + (2 - 1)] = 8$

$3 + 7 = 3 \times [7 + (3 - 1)] = 27$

$4 + 5 = 4 \times [5 + (4 - 1)] = 32$

$5 + 8 = 5 \times [8 + (5 - 1)] = 60$

$6 + 7 = 6 \times [7 + (6 - 1)] = 72$

Therefore,

$7 + 8 = 7 \times [8 + (7 - 1)] = 98$

$x + y = x[y + (x - 1)] = x^2 + xy - x$

91. DOMINO THEORY

Suppose you had two dominoes that looked as given below:

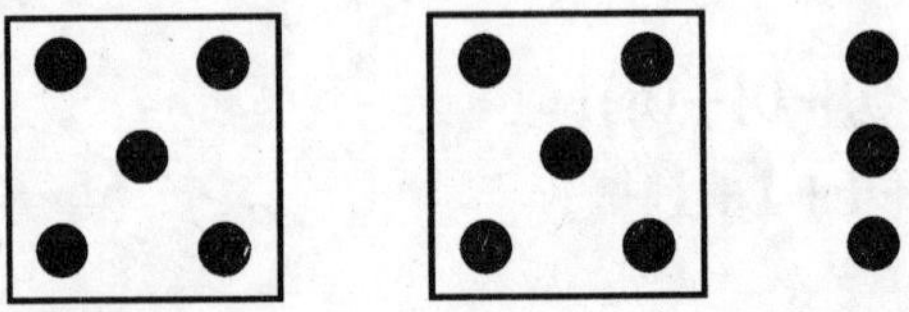

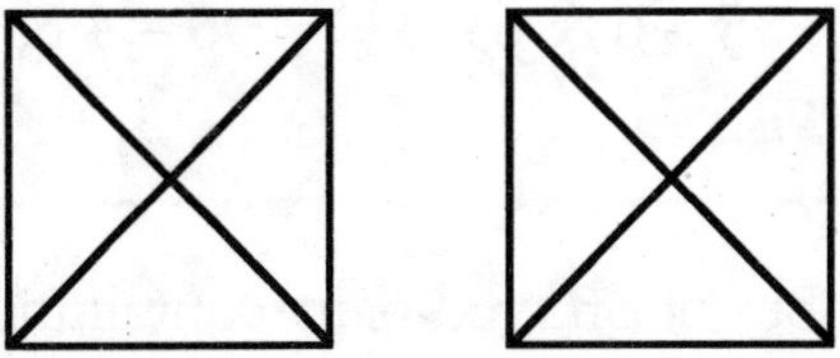

The X's bottom of the dominoes mean that these areas are not visible, so all you know is that one of the dominoes has a five-spot and the other has a six-spot. What is the probability that you can form an end-to-end chain of all 28 dominoes – with the two depicted dominoes in first and last position – subject to the usual rule that the number on the right of any domino in the chain always equals the number on the left of the next domino?

Answer

The probability is 17/48.

The key observation is that in order for a continuous chain of dominoes to be made in the manner described, the leftmost and rightmost spots must be the same.

Now, we make the assumption that for any pair of dominoes that share a spot, we can always construct a continuous chain with that spot at either end.

With that assumption under our belts, here is the complete set of ordered pairs that:

1. Contain at least one 5 and one 6 and 2
2. Contain a pair of repeated spots to serve as the endpoints of the chain:

5 – 0 & 6 – 0	5 – 6 & 6 – 0	5 – 0 & 6 – 5
5 – 1 & 6 – 1	5 – 6 & 6 – 1	5 – 1 & 6 – 5
5 – 2 & 6 – 2	5 – 6 & 6 – 2	5 – 2 & 6 – 5
5 – 3 & 6 – 3	5 – 6 & 6 – 3	5 – 3 & 6 – 5

5 – 4 & 6 – 4   5 – 6 & 6 – 4   5 – 4 & 6 – 5

5 – 5 & 6 – 5

5 – 6 & 6 – 6

The total number of ordered pairs containing at least one 5 and at least one 6 equals 48 (7 x 7 minus the 5 – 6 domino, which can't be repeated), so the desired probability must be 17/48.

92. COUNTERFEIT COIN

In front of you, there are 9 coins. They all look absolutely identical, but one of the coins is fake. However, you know that the fake coin is lighter than the rest, and in front of you is a balance scale. What is the least number of weightings you can use to find the counterfeit coin?

Answer

The answer is 2.

First, divide the coins into 3 equal piles. Place a pile on each side of the scale, leaving the remaining pile of 3 coins off the scale. If the scale does not tip, you know that the 6 coins on the scale are legitimate, and the counterfeit is in the pile in front of you. If the scale does tip, you know the counterfeit is in the pile on the side of the scale that rose up. Either way, put the 6 legitimate coins aside. Having only 3 coins left, put a coin on each side of the scale, leaving the third in front of you. The same process of elimination will find the counterfeit coin.

93. HOW COME?

Take 9 from 6, 10 from 9, and 50 from 40 leaves 6. How come?

Answer

SIX – 9 (IX) = S
9 (IX) – 10 (X) = I
40 (XL) – 50 (L) = X
⇒ SIX

94. MONK'S STEPS

A monk has a very specific ritual for climbing up the steps to the temple. First he climbs up to the middle step and meditates for 1 minute. Then he climbs up 8 steps and faces east until he hears a bird singing. Then he walks down 12 steps and picks up a pebble. He takes one step up and tosses the pebble over his left shoulder. Now, he walks up the remaining steps three at a time which only takes him 9 paces. How many steps are there?

Answer

There are 49 steps.

Since there is a middle step, the number of steps is odd. So the number of steps has the form 2N+1, where N is a number we have to figure out. Let's number the steps starting from the bottom, so the first step is 1, then the second is 2, and so on. The middle step is then the (N+1)st step, since it has N steps below it and N steps above it. Now let's just follow the monk with this notation in mind.

He goes to the middle step: that's step N + 1.

He climes 8 more: going to step (N + 1) + 8 = N + 9.

He walks down 12 steps: getting him to step (N + 9) - 12 = N - 3.

He takes one step up: so he's at (N - 3) + 1 = N - 2.

Nine paces of three steps each gets him to the top. In nine three-step paces we travel 9*3 = 27 steps. Since this gets us to the top step, which is step 2N + 1, we're told that (his current position) + 27 = (the top step) = 2N + 1.

Since his current position was N - 2 we find that N - 2 + 27 = 2N + 1.

Moving all the numbers to the left hand side and all the Ns to the right and simplifying we find that 24 = N. So there are 2N + 1 = 48 + 1 = 49 steps.

95. PARADOX PROBABILITY

Four balls are placed in a hat. One is white, one is blue and the other two are red. The bag is shaken and someone draws two balls from the hat. He looks at the two balls and announces that at least one of them is red. What are the chances that the other ball he has drawn out is also red?

Answer

There are six possible pairings of the two balls withdrawn,

RED	+	RED
RED	+	WHITE
WHITE	+	RED
RED	+	BLUE
BLUE	+	RED
WHITE	+	BLUE & BLUE + WHITE

We know that the WHITE + BLUE & BLUE +

WHITE combination have not been drawn. This leaves six possible combinations remaining. Therefore the chances that the RED + RED pairing has been drawn are 1 in 6. Many people cannot accept that the solution is not 1 in 3, and of course it would be, if the balls had been drawn out separately and the colour of the first ball announced as red before the second had been drawn out. However, as both balls had been drawn together, and then the colour of one of the balls announced, then the above solution, 1 in 6, must be the correct one.

96. LARGEST PRODUCT

Using the digits 1 up to 9, two numbers must be made. The product of these two numbers should be as large as possible. All digits must be used exactly once. Which are the requested two numbers?

Answer

The digits of the requested two numbers obviously form descending sequences. Furthermore, if you have two pairs of numbers with equal sums, the pair of which the numbers have the smallest absolute difference, is the one of which the numbers have the largest product. Using this knowledge, the two numbers can easily be constructed by placing the digits one by one, starting with 9 and ending with 1:

9

8 → 96

87 → 964

875 → 9642

8753 → 9642

87531

Hence, the requested two numbers are 9642 and 87531 (and the product of these two numbers is 843973902).

97. PAINTED CUBE

You have a 10 inch by 10 inch cube that is made up of little 1 inch by 1 inch cubes. You paint the outside of the big cube grey. How many of the little cubes get painted?

Answer

First, paint the top and bottom of the outside. The top is a 10x10 square, so 100 little cubes get painted. The bottom is a 10x10 square, so 100 little cubes get painted. That's 200 so far... Now, paint the front and back of the outside. The front is a 10x10 square. That's 100 little cubes. But, we've already counted the top stripe of cubes and the bottom strips since these got hit when we did the top and bottom. So, for the new little cubes we're hitting, we have an 8x10 grid. That's 80 more little cubes getting painted. So far: 100 + 100 + 80 + 80

Now, we do the left and right side of the big cube. Thinking about it the same way we did above, we'll be hitting an 8x8 set of new cubes with paint. That's 64 for the right side and 64 for the left side.

Total: 100 + 100 + 80 + 80 + 64 + 64 = 488

98. AMIT'S STRATEGY

Amit and Daksh are playing bets. Amit gives $10 to Daksh and Daksh deals four card out of a normal 52 card deck which is chose by him completely randomly. Daksh keeps them facing down and take the first card and show it to Amit. Amit have a choice of either keeping it or to look at the second card. When the second card is shown to him, he again has the choice of keeping or looking at the third which is followed by the third card as well; only if he does not want the third card, he will have to keep the fourth card.

If the card that is being chosen by Amit is n, Daksh will give him. Then the cards will be shuffled and the game will be played again and again. Now you might think that it all depends on chance, but Amit has come up with a strategy that will help him turn the favor in his odds. Can you deduce the strategy of Amit?

Answer

To keep odds in his favor, Amit must choose a card from the first three whenever he sees a 9 or higher card.

This is because the probability that all cards that are selected randomly by Daksh are below 9 is

$32 \times 23 \times 30 \times 29 / 52 \times 51 \times 50 \times 49 = 0.133.$

In such a case, you lose from \$2 to \$9 with equal probability = 0.133/8 = 0.0166

Let us now calculate the probability of the four cards being of value 9 or higher which will be equal to

$1 - 0.133 = 0.867.$

As Amit stops at the first sight of 9 or a higher card, he can possibly win -1, 0, 1, 2, 3 with an equal probability of

$0.867/5 = 0.173$

This will give him an overall expected winning amount of 0.14 per game he plays.

Also note that if Amit decides to stop at a card 10 or higher, the expected winning amount will be 0.09 per game. It can be a strategy to win more but will not stand quite effective.

With any other choice of stopping, Amit will be having negative chances of winning.

99. COUNT THE TRIANGLES

Can you count the number of triangle on the picture given below?

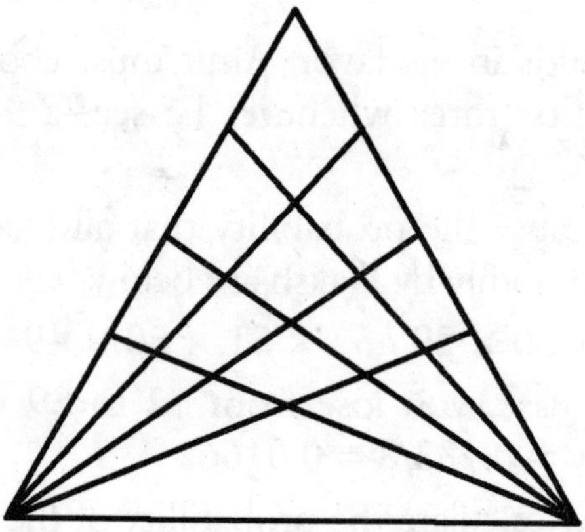

Answer

64 triangles

100. Marble Time

Anmol had a bag containing colourful marbles. The colours

are white, grey and black. The total number of marbles he had in his bag is 60. There are 4 times as many white marbles as grey marbles. 6 more black marbles than grey marbles. How many marbles of each colour did Anmol have?

Answer

Let's take W as white, G as grey and B as black.

W + G + B = 60

As there are 4 times white marbles than grey marbles, then

W = 4G

There are 6 more black marbles than grey marbles, so

B = G + 6

Therefore,

W + G + B = 60

4G + G + G + 6 = 60

6G + 6 = 60

6G = 60 - 6

6G = 54

G = 54/6 = 9

W = 4G = 4 x 9 = 36

B = G + 6 = 9 + 6 =15

Finally, there are 9 grey marbles, 36 white marbles and 15 black marbles.

101. 100 GRASSHOPPERS ON A TRIANGULAR BOARD

A triangular board has been cut into 100 small triangular cells by the lines parallel to its sides. Two cells that share a side are said to be neighbours. All at once, the grasshoppers hop from their cells to neighbouring cells. This happens 9 times. Prove that at least 10 cells are now empty.

Answer

The board is naturally coloured into two colours so that the neighbouring cells are coloured differently. Let the colours be black and grey and label the grasshoppers by the colour of their cells. Convince yourself that there are 55 black and 45 grey cells. In one hop, the grey grasshoppers move to the black cells thus emptying 55 cells. On the same hop, the grey grasshoppers move to the black cells thus filling at most 45 black cells.

Inevitably, at least 10 black cells will remain empty.

The next time the grasshoppers move, they can move into their original cells, thus filling all of them. The situation will return to the original one, while the argument will apply to every odd hop. Perhaps there may happen a greater mix-up of the grey and black hoppers, but the best that may be claimed is that after every odd hop, there are at least 10 empty cells.

In general, there are n(n+1)/2 grey cells and n(n-1)/2 black cells. The difference being n(n+1)/2 - n(n-1)/2 = n, there are always at least n empty cells after an odd number of hops.

The specific number of hops (9) is nothing but a grey herring.

102. WEIGH

Jatin weighs half as much as Vikram, and Yuvraj weighs three times as much as Jatin. Together, they weigh 720 pounds. How much does each man weigh?

Answer

If Jatin weighs twice as much as Vikram, and Yuvraj three times as much, dividing their total weight by six gives us Jatin 's weight (think $x + 2x + 3x = 720$). 720 divided by 6 tells us that Jatin weighs 120 pounds, so Vikram weighs 240 pounds and Yuvraj weighs 360 pounds.

103. HOW MANY CARTONS?

A merchant can place 8 large boxes or 10 small boxes into a carton for shipping. In one shipment, he sent a total of

96 boxes. If there are more large boxes than small boxes, how many cartons did he ship?

Answer

There are 11 cartons in total.

7 large boxes ($7 \times 8 = 56$ boxes)

4 small boxes ($4 \times 10 = 40$ boxes)

Total 96 boxes were shipped

104. HAT AND RIVER PROBLEM

Joe puts his canoe in the river and starts paddling upstream. After a mile his hat falls in the river. Ten minutes later he realises his hat is missing and immediately paddles downstream to retrieve it. He catches up to it at the same place he launched his canoe in the first place. What is the speed of the river current?

Answer

Let,

p = paddling speed in still water

c = current of the water

d = distance Joe paddled upstream since his hat fell out of the canoe

Recall that distance = rate × time

Consider the unknown distance Joe went upstream since his hat fell out:

(1) $d = (p - c) \times (1/6)$

We're also given that it took the same time for the hat to float downriver a mile as it took for Joe to travel d miles upstream and 1+ d miles downstream. Let's set up an equation, balancing for time.

(2) Time for hat to float one mile downstream = Time for Joe to travel d miles upstream + Time for Joe to travel 1 + d times downstream.

Time for hat to float one mile downstream = one mile/ rate of current = $1/c$

Time for Joe to travel d miles upstream = 1/6 (in hours), as provided in the question.

Time for Joe to travel 1+d times downstream = distance Joe traveled downstream / rate Joe traveled downstream = $[1 + (1/6) \times (p - c)]/(p + c)$.

The distance is the one mile the hat floated in the river + the $(1/6) \times (p - c)$ from equation (1). Joe's speed going downstream is $p + c$, which is the sum of his paddling speed and current speed.

Now, we're ready to solve for equation (2)

$1/c = (1/6) + [1 + (1/6) \times (p - c)]/(p + c)$

$1/c = (1/6) + (6+p-c)/(6p + 6c)$

$6/c = 1 + (6 + p - c)/(p + c)$ Multiplying both sides by 6

$6/c = (p + c)/(p + c) + (6 + p - c)/(p + c)$

$6/c = (6 + 2p)/(p + c)$

$6p + 6c = 6c + 2pc$

$6p = 2pc$

$3 = c$

So, the current is 3 miles per hour.

105. HAT AND RIVER PROBLEM

Three players play a game where the loser must double the money of both the other two players. They play this game three times and each player loses once. After three rounds each player has 24. How much did each player start with?

Answer

Let's call the players P1, P2, and P3, and the starting bankrolls as follow:

P1: x

P2: y

P3: z

Assume that P1 loses the first round. After he pays the winners, the bankrolls will be:

P1: $x - y - z$

P2: $2y$

P3: $2z$

Assume that P2 loses the second round. After he pays the winners, the bankrolls will be:

P1: $2x - 2y - 2z$

P2: $-x + 3y - z$

P3: $4z$

Assume that P3 loses the third round. After he pays the winners, the bankrolls will be:

P1: $4x - 4y - 4z$

P2: $-2x + 6y - 2z$

P3: $-x - y + 7z$

All three of these sums must equal 24. So, we have

three equations and three unknowns. That is enough to do a some simple matrix algebra to get x = 39, y = 21, z = 12.

106. GUESS THE AGE

Person x and y have the following conversation:

x: I forgot how old your three kids are.

y: The product of their ages is 36.

x: I still don't know their ages.

y: The sum of their ages is the same as your house number.

x: I still don't know their ages.

y: The oldest one has red hair.

x: Now I know their ages!

How old are they?

Answer

From the statement that the product of their ages is 36 the possibilities of the three individual ages are:

1, 1, 36

1, 2, 18

1, 3, 12

1, 4, 9

1, 6, 6

2, 2, 9

2, 3, 6

3, 3, 4

From the statement that the sum equals the house

number it is possible to eliminate all but two possibilities. The sums of the rest two possibilities are unique and would allow for an immediate answer. For example if the house number were 16 the ages must be 1, 3, and 12 which is not possible. The two remaining possibilities are 2, 2, and 9; or 1, 6, and 6.

After the clue that the oldest has red hair you can eliminate 1, 6, and 6 because the oldest two have the same age thus there is no oldest son. The only remaining possibility is 2, 2, and 9.

107. MEETING AT THE RESTAURANT

Two people arrive in a restaurant independently. Each arrives a random time between 5pm and 6pm, distributed uniformaly (no moment in this range is any more likely for arrival than another). What is the probability they arrived within 10 minutes of each other?

Answer

Draw a 6 by 6 square. Plot the arrival time of one person vertically. Plot the arrival time of the other person horizontally. The times they arrive within 10 minutes of each other can be represented by a diagonal stripe across the square. The area of the entire square is 36. The area of the portions outside the stripe is 25. Thus, the area of the stripe is 11. Considering the uniformity of the arrival times the probability of arriving within ten minutes is the area of the stripe divided by the area of the square = 11/36.

108. ZEROS AND ONES

What is the smallest integer greater than 0 that can be

written entirely with zeros and ones and is evenly divisible by 225?

Answer

The prime factorization of 225 is $5 \times 5 \times 3 \times 3$. So the answer will be both a multiple of 25 and of 9.

All multiples of 25 end in 00, 25, 50, or 75. The only one of these composed of 0's and 1's is obviously 00, so the answer must end in 00. The hard part is finding a series of 0's and 1's preceding the 00 that will make the entire number divisible by 9.

If you didn't already know the following trick then this problem would be very hard. If you did know it then the problem was likely very easy. The trick is that if the sum of digits of a number is divisible by 9 then the number itself is also divisible by 9. Note that this is true for 3 also. For example the number 17685 is divisible by 9 because $1 + 7 + 6 + 8 + 5 = 27$, and 27 is divisible by 9.

To prove this let's consider any five digit number, abcde. This number can be expressed as follows.

$a \times 10000 + b \times 1000 + c \times 100 + d \times 10 + e =$

$a \times (9999+1) + b \times (999+1) + c \times (99+1) + d \times (9+1) + e \times 1 = a \times 9999 + b \times 999 + c \times 99 + d \times 9 + a + b + c + d + e$

$= 9 \times (a \times 1111 + b \times 111 + c \times 11 + d \times 1) + a + b + c + d + e$

Thus, $9 \times (a \times 1111 + b \times 111 + c \times 11 + d \times 1)$ is a multiple of 9.

So, if $a + b + c + d + e$ is also a multiple of 9 then the entire number must be a multiple of 9. Note also that the remainder of abcde/9 is the same as the remainder of (a + b + c + d + e)/9.

The smallest number consisting of all 1's and divisible by 9 is thus 111,111,111. Now, add the two zeros at the end results in the answer to the problem: 11,111,111,100

109. POTHOLE PROBLEM

Any given length of highway is equally likely to have as many potholes as any other length of equal size. The average number of potholes per mile of highway is 3. What is the probability that 2 miles of highway have 3 or fewer potholes?

Answer

The number of potholes of mile of highway has a Poisson distribution with a mean of 3. The number of potholes in two miles of highway would have a Poisson distribution with a mean of 6. The probability of x potholes in two miles of highway is $e^{-6} \times 6^x/x!$

The probability of 0 potholes is $e^{-6} \times 6^0/0! = e^{-6}$

The probability of 1 pothole is $e^{-6} \times 6^1/1! = 6 \times e^{-6}$

The probability of 2 potholes is $e^{-6} \times 6^2/2! = 18 \times e^{-6}$

The probability of 3 potholes is $e^{-6} \times 6^3/3! = 36 \times e^{-6}$

Take the sum the answer is $61 \times e^{-6} = \sim 0.151204$

110. OVERLAP AGAIN

I have a clock (12 hour format) and both the needles of clock overlaps at 12:00. After how much time, they will overlap again?

Answer

The clock will overlap again after 1hr 5 min at 01:05.

111. INSURANCE COMPANY

An insurance company issues policies to two classes of people, as shown in the table below.

Problem 153

Class	Probability of Death	Death Benefit	Number in Class
A	.01	200	500
B	.05	100	300

The company charges each customer the product of expected claim amount and a constant k. How much should k be so that the probability of total claims exceeding total revenue is 5%?

Answer

Let a be an individual in class A and b be an individual in class B.

$E(a) = 200 \times .01 = 2$

$E(a^2) = 2002 \times .01 = 400$

$Var(a) = E(a^2) - (E(a))^2 = 400 - 4 = 396$

$E(b) = 100 \times .05 = 5$

$E(b^2) = 1002 \times .05 = 500$

$Var(b) = E(b^2) - (E(b))^2 = 500 - 25 = 475$

The variance of total claims is $500 \times 396 + 300 \times 475 = 340{,}500$

The standard deviation of total claims is $3405001/2 = \sim 583.52$

Expected total claims is $500 \times 2 + 300 \times 5 = 2500$

Pr(total claims <= total revenue) = .95

Pr(C <= R) = .95 (where C = total claims, R = total revenue)

Pr(C <= R) = .95

Pr(C – m <= R - m) = .95 (where m is the expected total claims)

Pr(C-2500 <= R-2500) = .95

Pr((C-2500)/s <= (R-2500)/s) = .95 (where s is the standard deviation)

Pr((C-2500)/583.52 <= (R-2500)/583.25) = .95

We can now use the central limit theorem because (C-2500)/583.52 has a normal distribution with mean of 0 and standard deviation of 1.

(R-2500)/583.25 = 1.645

R = 3459.89

So, the insurance company needs 3459.89 in revenue. The expected cost of claims is 2500.

So, k = 3459.89/2500 = 1.38

112. LARGE TANK

A large tank has a steadily flowing intake and 10 outlet valves, the latter being all of the same size. With 10 outlets open, it takes two and one half hours to empty the tank; with 6 outlets open it takes five and one half hours to empty the tank. After the tank is empty and with all 10 outlets closed, how long will it take to fill the tank?

Answer

Let x be the input rate in gallons per hour into the tank.

Let y be the output rate in gallons per hour from the tank for each value.

Let w be the number of gallons in the tank.

We know that it takes 2.5 hours to empty the tank

with 10 valves open. Over the 2.5 hours the sum of the water in the tank initially and the water going into the tank will equal the amount of water leaving the tank. Let's set that up as an equation.

$w + 2.5x = 2.5 \times 10y$

We also know that it takes 5.5 hours to empty the tank with 6 valves open. Let's set that up as an equation.

$w + 5.5x = 5.5 \times 6y$

I don't like decimals so let's multiply both equations by 2:

$2w + 5x = 50y$

$2w + 11x = 66y$

To solve the problem we need to know the relationship between w and x, so let's solve for y in the first equation and substitute in the second:

$y = (2w + 5x)/50$

Substituting this in the second equation...

$2w + 11x = 66 \times (2w + 5x)/50$

$100w + 550x = 132w + 330x$

$220x = 32w$

$w = 220x/32$

$w = 6.875x$

So, 6.875 times the input rate equals the capacity of the tank. Thus, it would take 6.875 hours to fill the tank.

113. THREE CARDS

Three playing cards, removed from an ordinary deck, lie face down in a horizontal row. Immediately to the right of the King there's a Queen or two. Immediately to the left of a Queen there's a Queen or two. Immediately to the left of a Heart there's a Spade or two. Immediately to the right of a Spade there's a Spade or two. Name the three cards in order.

Answer

Let's label the clues as follows:

To the right of the King there's a Queen or two.

To the left of a Queen there's a Queen or two.

To the left of a Heart there's a Spade or two.

To the right of a Spade there's a Spade or two.

From clues 1 and 2 we can infer that a king is on the left and a queen on the right and the middle card is either a queen or 2.

First consider the possibility that a 2 is in the middle. If there are two consecutive spades (from clue 4) then then three cards must be the king of spades, 2 of spades, queen of hearts. If to the right of a spade there is a 2 (again from rule 4) then the three cards are the kings of spades, a two of unknown suit, and the queen of hearts. So, if we assume a 2 in the middle it is not certain what the suit of the 2 is.

Next consider the possibility that a queen is in the

middle. Clue 4 tells us there must be two consecutive spades, since there is no 2. Both queens cannot be spades since they are from the same deck, thus the left and center cards must be spades. Clue 3 tells us the right queen is a heart. So, the three cards would be the king of spades, queen of spades, and queen of hearts.

So, a 2 in the middle does not lead to a definitive answer, but a queen in the middle does. The question implies there is just one possible answer. Therefore, there is a queen in the middle, leading to the final answer of king of spades, queen of spades, and queen of hearts.

114. BEHIND THE DOOR

On a game show there are three doors. Behind one door is a new car and behind the other two are goats. Every time the game is played the contestant first picks a door. Then the host will open one of the other two doors and always reveals a goat. Then the host gives the player the option to switch to the other unopened door. Should the player switch?

Answer

The key to this problem is that the host is predestined to open a door with a goat. He knows which door has the car so regardless of which door the player picks he always can reveal a goat behind another door.

Let's assume that the prize is behind door 1. Following are what would happen if the player had a strategy of not switching.

Player picks door 1 → player wins

Player picks door 2 → player loses

Player picks door 3 → player loses

Following are what would happen if the player had a strategy of switching.

Player picks door 1 → Host reveals goat behind door 2 or 3 → player switches to other door → player loses

Player picks door 2 → Host reveals goat behind door 3 → player switches to door 1 → player wins

Player picks door 3 → Host reveals goat behind door 2 → player switches to door 1 → player wins

So by not switching the player has 1/3 chance of winning. By switching the player has a 2/3 chance of winning. So the player should definitely switch.

The plain simple English reason the probability of winning increases to 66.67% by switching, lays in the fact that the host always reveals a goat. If the host didn't know which door had the car, then the probability of having a win would go up to 50% after revealing one of the goats. Such is the case on the show "Deal or no Deal." On that show, the host does not know where the million dollar case is. So, as cases are eliminated that do not have the million dollars, the probability increases that every remaining case has it, equally. When there are only two cases left and two prizes, each case has a 50% chance of each prize.

115. NINE MINUTE EGG

You are a cook in a remote area with no clocks or other way of keeping time other than a 4 minute hourglass and a 7 minute hourglass. You do have a stove however with water in a pot already boiling. Somebody asks you for a 9 minute egg, and you know this person is a perfectionist and will be able to tell if you undercook or overcook the eggs by even a few seconds. What is the least amount of time it will take to prepare the egg?

Answer

Solution

Time	**Action**	**Time remaining before action**		**Time remaining after action**	
		4 Minute	**7 Minute**	**4 Minute**	**7 Minute**
0	Turn over both hourglasses	0	0	4	7
4	Turn over 4 minute hourglass	0	3	4	3
7	Turn over 7 minute hourglass	1	0	1	7

8	Turn over 7 minute hourglass	0	6	0	1
9	Take egg off stove	0	0	0	0

116. FURNITURE FACTORY

A factory that produces tables and chairs is equipped with 10 saws, 6 lathes, and 18 sanding machines. It takes a chair 10 minutes on a saw, 5 minutes on a lathe, and 5 minutes of sanding to be completed. It takes a table 5 minutes on a saw, 5 minutes on a lathe, and 20 minutes of sanding to be completed. A chair sells for 10 thousand and a table sells for 20 thousand. How many tables and chairs should the factory produce per hour to yield the highest revenue, and what is that revenue?

Answer

The 10 saws can produce 600 minutes of work per hour (10 saws × 60 minutes).

The 6 lathes can produce 360 minutes of work per hour (6 lathes × 60 minutes).

The 18 sanding machines can produce 1080 minutes of work per hour (18 sanding machines × 60 minutes).

Let c be the number of chairs produced per hour and t the number of tables produced per hour.

The number of saws limit the combination of chairs and tables to $600 = 10c + 5t$.

The number of lathes limit the combination of chairs and tables to $360 = 5c + 5t$.

The number of sanding machines limit the combination of chairs and tables to 1080 = 5c + 20t.

Next graph these three lines. It should be expected that the answer will lie on the intersection of two of these lines or to make all chairs or all tables. The intersection of the saw and sanding machine line occurs outside of how many chairs the lathe can make so this combination is not a viable answer. The saw and lathe lines cross at 48 chairs and 24 tables. The lathe and sanding machine lines cross at 24 chairs and 48 tables.

Next determine the revenue at all points of intersection.

Chairs	Tables	Revenue
60	0	600
48	24	960
24	48	1200
0	54	1080

So, the optimal answer is to make 24 chairs and 48 tables for revenue of 1200 thousand per hour.

To check it will take 24 × 10 + 48 × 5 = 480 minutes of saw time. There are 600 minutes available so the saws will be idle 20% of the time.

It will take 24 × 5 + 48 × 5 = 360 minutes of lathe time which is exactly what we have.

It will take 24 × 5 + 48 × 20 = 1080 minutes of sanding machine time which is exactly what we have.

117. LAND AND SEA

You are in a race in which the starting line is at a certain point on a straight beach. The finish line is in the water. One way to arrive at the finish line is to run 4 kilometers down the beach, make a 90 degree turn and swim 1 kilometer. However, you may cut into the water at any point. You speed on land is 6 km/h and you speed in water is 2 km/h. At what point, measured from the starting line, should you cut into the water?

Answer

Let y be the solution. Let x = 4 - y.

The time to reach the finish line is:

$$T = \frac{(4 - x)}{6} + \frac{(1 + x^2)^{1/2}}{2}.$$

Set the derivative equal to 0:

$$dt/dx = -1/6 + 1/2 \times 1/2 \times 2x \times (1 + x^2)^{-1/2} = 0$$

The solution is $x = \left(\frac{1}{8}\right)^{1/2}$

y = 4 - x

118. TEACHER'S CHALK

Your teacher has a total of 9 chalks. When a chalk reduces to 1/3 of its original size, it gets too small for her to hold for writing and hence she keeps it aside. But your teacher hates wasting things and so, when she realizes that she has enough of these small pieces to join and make another chalk of the same size, she joins them and uses the new chalkstick. If she uses one chalk each day, how many days would the 9 chalks last?

Answer

Your teacher uses one chalk each day. Hence the total number of days she uses 9 chalks is 9. Each chalk leaves a fraction of 1/3 its size... so 9 such fractions remain. Since 3 such fractions are joined to give a new chalk, your teacher combines all the fractions to get 3 chalks which can again be used for 3 days. Hence, she has managed to use 9 chalks for (9 + 3) days!

But, what about the leftovers of the chalks used over the last 3 days?? They can be joined to form yet another chalk... which means another day! So, your teacher uses the 9 chalks for a total of 13 days.

119. FIVE SOCKS

In a drawer are two red socks and three blue socks. A sock is drawn at random from the drawer, with replacement, one million times. What is the range, with the expected outcome as the midpoint of the range, such that the probability is 95% that the number of red socks drawn falls within this range?

Answer

The mean number of reds is $1{,}000{,}000 \times 0.4 = 400{,}000$.

The standard deviation of the number of reds is sqr($1{,}000{,}000 \times 0.4 \times (1 - .4)$) = 489.9

Let l denote the lower bound of the range, let u denote the upper bound, and let x denote the number of reds drawn:

Pr ($l <= x <= u$) = .95

Pr ($l - 400{,}000 <= x - 400{,}000 <= u - 400{,}000$) = .95

Pr ($(l - .5 - 400{,}000)/489.9 <= Z <= (u + .5 - 400{,}000)/489.9$) = .95 where Z denotes a random variable distributed according to the standard normal distribution.

Next we want the probability that the number of reds will fall on either side of this range to be .025 per side.

Pr($Z <= (u + .5 - 400{,}000)/489.9$) = .975

$(u + .5 - 400{,}000)/489.9 = 1.96$

$u - 399{,}999.5 = 960.2$

$u = 400{,}960$

$Pr(Z <= (l - 400{,}000.5)/489.9) = .025$

$(l - 400{,}000.5)/489.9 = -1.96$

$l = 399{,}040$

So, the range is 399,040 to 400,960.

120. MOUSE RUBIK CUBE

A cubic piece of cheese has been subdivided into 27 sub cubes (so that it looks like a Rubik's Cube). A mouse starts to eat a corner sub cube. After eating any given sub cube it goes on to another adjacent sub cube. Is it possible for the mouse to eat all 27 sub cubes and finish with the center cube?

Answer

Consider the sub cubes to form a three dimensional checkerboard composed of sub cubes of Swiss and cheddar cheese. Assume the corners and centers of each face are Swiss cheese and the rest of the sub cubes, including the center, are cheddar cheese. Assume the mouse eats two sub cubes an hour. At the beginning of every hour the mouse will be starting on a Swiss sub cube and will end the hour with a sub cube of cheddar cheese. However the center sub cube is also cheddar. Since two adjacent cubes must be of different kinds of cheese the mouse cannot eat the center cube last.

121. DROPPING A BILLIARD BALL

It is your task to determine how high you can drop a

billiard ball without breaking it. There is a 100 story building and you must determine the highest floor from which you can drop a ball without breaking it. You have only two billiard balls to use as test objects, if both of them break before you determine the answer; you have failed at your task. How can you determine the breaking point in which the maximum necessary dropping is at a minimum?

Answer

First drop the first ball from the 14th floor. If it breaks you can determine the exact breaking point with the other ball in at most 13 more droppings, starting at the bottom and going up one floor at a time.

If the first ball survives the 14 floor drop then drop it again from the 27th (14 + 13) floor. If it breaks you can determine the exact breaking point with at most 12 more droppings.

If the first ball survives the 27 floor drop then drop it again from the 39th (14 + 13 + 12) floor. If it breaks you can determine the exact breaking point with at most 11 more droppings.

Keep repeating this process always going up one less floor than the last dropping until the first ball breaks. If it breaks on the xth dropping you will only need at most 14 - x more droppings with the second ball to find the breaking point. By the 11th dropping of the first ball, i.e. Repeating the process 11 times (14+13+12+11+10+9+8+7+6+5 +4)if you get that far, you will have reached the 99th floor.

122. COCONUTS AND MONKEYS

Ten people land on a deserted island. There they find lots of coconuts and a monkey. During their first day they gather coconuts and put them all in a community pile. After working all day they decide to sleep and divide them into two equal piles the next morning. That night one castaway wakes up hungry and decides to take his share early. After dividing up the coconuts he finds he is one coconut short of ten equal piles. He also notices the monkey holding one more coconut. So he tries to take the monkey's coconut to have a total evenly divisible by 10. However when he tries to take it the monkey conks him on the head with it and kills him. Later another castaway wakes up hungry and decides to take his share early. On the way to the coconuts he finds the body of the first castaway, which pleases him because he will now be entitled to 1/9 of the total pile. After dividing them up into nine piles he is again one coconut short and tries to take the monkey's coconut. Again, the monkey conks the man on the head and kills him. One by one each of the remaining castaways goes through the same process, until the 10th person to wake up gets the entire pile for himself. What is the smallest number of possible coconuts in the pile, not counting the monkeys?

Answer

We know from the first man that the total must be one less than a number divisible by 10.

We know from the second man that that total must be one less than a number divisible by 9.

After considering all 10 people the final number must

be one less than a number divisible by every number from 1 to 10.

Certainly 10! = 3628800 works but there is smaller solutions.

The smallest answer is LCM (1, 2, 3, 4, 5, 6, 7, 8, 9, 10) -1

$= 7 \times 2 \times 2 \times 2 \times 3 \times 3 \times 5 - 1 = 2519$

123. SOLVE IT

ABCDE × 4 = EDCBA. Solve for A, B, C, D, and E where each is a unique integer from 0 to 9.

Answer

It is obvious that A can be no more than 2. If A were 3 then 3BCDE × 4 would be at least 120,000 which is more than five digits. Also A must be an even number because EDCBA is an even number since it is the product of at least one even number (4). We can eliminate A=0 because E would have to be 5 (5 × 4=0) but BCDE × 4 could not hope to reach 50,000. So, A must be 2.

Next consider E. E × 4 must end in the digit 2. The only numbers that work for are 3 and 8. However with A = 2 EDCBA must be at least 80,000. So, 8 is the only number that satisfies both conditions.

Next consider B. We already know that 2BCD8 × 4 is at least 80000 and less than 90000. B cannot be more than 2 because then 2BCD8 × 4 would be more than 80000. 2 is already taken so, B must be 0 or 1.

Let's consider the case that B = 0. Then D8 × 4 must

end in the digit 02. However, there is no D that satisfies this condition. So, B must be 1.

Next consider D. D8 × 4 must end in the digits 12. The only possibility is D=7 (78 × 4 = 312).

Now solve for C:

21C78 × 4 = 87C12

84312 + 400C = 87012 + 100C

2700 = 300C

C = 2700/300 = 9

So, ABCDE = 21978

124. ANTS ON THE BOARD

There are 100 ants on a board that is 1 meter long, each facing either left or right and walking at a pace of 1 meter per minute. The board is so narrow that the ants cannot pass each other; when two ants walk into each other, they each instantly turn around and continue walking in the opposite direction. When an ant reaches the end of the board, it falls off the edge. From the moment the ants start walking, what is the longest amount of time that could pass before all the ants have fallen off the plank? You can assume that each ant has infinitely small length.

Answer

The longest amount of time that could pass would be 1 minute.

If you were looking at the board from the side and could only see the silhouettes of the board and the ants, then when two ants walked into each other and turned around, it would look to you as if the ants had walked right by each other.

So we can treat the board as if the ants are walking past each other. In this case, the longest any ant can be on the board is 1 minute (since the board is 1 meter long and the ants walk at 1 meter per minute). Thus, after 1 minute, all the ants will be off the board.

125. PLAYING WITH RINGS

A man comes to a small hotel where he wishes to stay for 7 nights. He reaches into his pockets and realizes that he has no money, and the only item he has to offer is a gold chain, which consists of 7 rings connected in a row (not in a loop).

The hotel proprietor tells the man that it will cost 1 ring per night, which will add up to all 7 rings for the 7 nights.

"Ok," the man says. "I'll give you all 7 rings right now to pre-pay for my stay."

"No," the proprietor says. "I don't like to be in other people's debt, so I cannot accept all the rings up front."

"Alright," the man responds. "I'll wait until after the seventh night, and then give you all of the rings."

"No," the proprietor says again. "I don't like to ever be owed anything. You'll need to make sure you've paid me the exact correct amount after each night."

The man thinks for a minute, and then says "I'll just cut each of my rings off of the chain, and then give you one each night."

"I do not want cut rings," the proprietor says. "However, I'm willing to let you cut one of the rings if you must."

The man thinks for a few minutes and then figures out a

way to abide by the proprietor's rules and stay the 7 nights in the hotel. What is his plan?

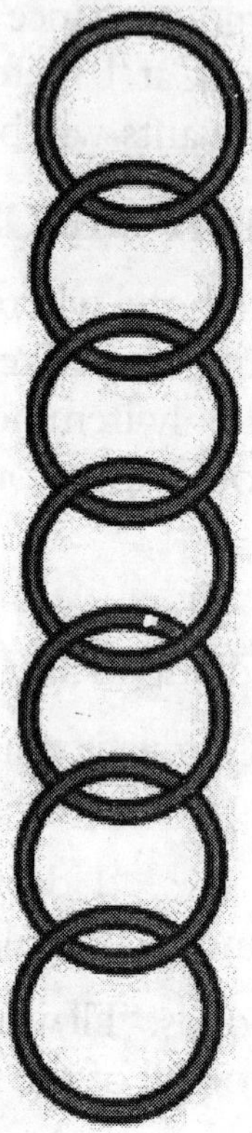

Answer

The man cuts the ring that is third away from the end of the chain. This leaves him with 3 smaller chains of length 1, 2, and 4. Then, he gives rings to the proprietor as follows:

After night 1, give the proprietor the single ring

After night 2, take the single ring back and give the proprietor the 2-ring chain.

After night 3, give the proprietor the single ring,

totalling 3 rings with the proprietor.

After night 4, take back the single ring and the 2-ring chain, and give the proprietor the 4-ring chain.

After night 5, give the proprietor the single ring, totalling 5 rings with the proprietor.

After night 6, take back the single ring and give the proprietor the 2-ring chain, totalling 6 rings with the proprietor.

After night 7, give the proprietor the single ring, totalling 7 rings with the proprietor.

126. AIRPLANE SEATS

People are waiting in line to board a 100-seat airplane. Steve is the first person in the line. He gets on the plane but suddenly can't remember what his seat number is, so he picks a seat at random. After that, each person who gets on the plane sits in their assigned seat if it's available, otherwise they will choose an open seat at random to sit in. The flight is full and you are last in line. What is the probability that you get to sit in your assigned seat?

Answer

There is a 1/2 chance that you'll get to sit in your assigned seat.

A common way to try to solve this riddle is to try to mathematically determine the chance that each person sits in your seat as they get on the plane. However, this math gets complicated quickly, and we can solve this riddle with a more analytical approach.

We first make two observations:

1. If any of the first 99 people sit in your seat, you

WILL NOT get to sit in your own seat.

2. If any of the first 99 people sit in Steve's seat, you WILL get to sit in your seat. To see why, let's say, for the sake of example, that Steve sat in A's seat, then A sat in B's seat, then B sat in C's seat, and finally, C was the person who sat in Steve's seat. We can see that this forms a sort of loop in which every person who didn't sit in their own seat is actually sitting in the seat of the next person in the loop. This loop will always be formed when a person finally sits in Steve's seat (and if Steve sits in his own seat, we would consider this to be a loop of length 1), and so after that point, everybody gets to sit in their own seat.

Based on these observations, we know that the instant that a passenger sits in either Steve's seat or your seat, the game for you is "over", and it is fully decided if you will be sitting in your seat or not.

Our final observation is that for each of the first 99 people, it is EQUALLY LIKELY that they will sit in Steve's seat or your seat. For example, consider Steve himself. There is a 1/100 chance that he will sit in his own seat, and a 1/100 chance that he'll sit in your seat.

So because there is always an equal chance of a person sitting in your seat or Steve's seat (and one of these situations is guaranteed to happen within the first 99 people), then there is an equal chance that you will or will not get your seat. So the chance you get to sit in your seat is 50%.

127. Bicycles and Tricycles

Last weekend, I went to play in the nearby park. It was

real fun! I rode my new bicycle that Mom had gifted me on my birthday.

On reaching the park, I saw that there were a total of 11 bicycles and tricycles. If the total number of wheels was 26, how many tricycles were there?

Answer

Assuming 2 wheels for each cycle, 11 cycles will have 22 wheels. But, there are 26 - 22 = 4 extra wheels.

As bicycles have 2 wheels and tricycles have 3 wheels, there is 1 extra wheel per tricycle in the park. Thus, the 4 extra wheels belong to 4 tricycles.

128. Wooden Block

A block of wood in the form of a cuboid 5" × 7" × 14" has all its six faces painted pink. If the wooden block is cut into 490 cubes of 1" × 1" × 1", how many of these would have pink paint on them?

Answer

The 1" × 1" × 1" cubes that do not have any pink paint on them will be at the core of the wooden block. This core will be 3" × 5" × 12", and will contain 180 cubes.

Out of a total of 490 cubes, there are 180 cubes without any pink paint on them. Therefore, the remaining 310 cubes will have one, two or three sides with pink paint on them (depending on whether they were at the face, edge or corner of the wooden block).

129. Two shepherds

A hunter met two shepherds, one of whom had three loaves and the other, five loaves. All the loaves were the same size. The three men agreed to share the eight loaves equally between them. After they had eaten, the hunter gave the shepherds eight bronze coins as payment for his meal. How should the two shepherds fairly divide this money?

Answer

The shepherd who had three loaves should get one coin and the shepherd who had five loaves should get seven coins. If there were eight loaves and three men, each man ate two and two-thirds loaves. So the first shepherd gave the hunter one-third of a loaf and the second shepherd gave the hunter two and one-third loaves. The shepherd who gave one-third of a loaf should get one coin and the one who gave seven-thirds of a loaf should get seven coins.

130. Always & Never

It's always 1 to 6,

it's always 15 to 20,

it's always 5,

but it's never 21,

unless it's flying.

What is that?

Answer

The answer is: a dice. An explanation: "It's always 1 to 6": the numbers on the faces of the dice, "it's always 15 to 20": the sum of the exposed faces when the dice comes to rest after being thrown, "it's always 5": the number of exposed faces when the dice is at rest, "but it's never 21": The sum of the exposed faces is never 21 when the dice is at rest, "unless it's flying": the sum of all exposed faces when the dice is flying is 21 (1 + 2 + 3 + 4 + 5 + 6).

131. Mirroring Clock

A boy leaves home in the morning to go to school. At the moment he leaves the house he looks at the clock in the mirror. The clock has no number indication and for this reason the boy makes a mistake in interpreting the time (mirror-image). Just assuming the clock must be out of order, the boy cycles to school, where he arrives after twenty minutes. At that moment the clock at school shows a time that is two and a half hours later than the time that the boy saw on the clock at home. At what time, the boy reaches school?

Answer

The difference between the real time and the time of the mirror image is two hours and ten minutes (two and a half hours, minus the twenty minutes of cycling). Therefore, the original time on the clock at home that morning could only have been five minutes past seven: The difference between these clocks is exactly 2 hours and ten minutes (note that also five minutes past one can be mirrored in a similar way, but this is not in the morning!). Conclusion: The boy reaches school at five minutes past seven plus twenty minutes of cycling, which is twenty-five minutes past seven!

132. Marbles

You are given a set of scales and 12 marbles. The scales are of the old balance variety. That is, a small dish hangs from each end of a rod that is balanced in the middle. The device enables you to conclude either that the contents of the dishes weigh the same or that the dish that falls lower has heavier contents than the other. The 12 marbles appear to be identical. In fact, 11 of them are identical, and one is of a different weight. Your task is to identify the unusual marble and discard it. You are allowed to use the scales three times if you wish, but no more. Note that the unusual marble may be heavier or lighter than the others. You are asked to both identify it and determine whether it is heavy or light.

Answer

Most people seem to think that the thing to do is weigh six coins against six coins, but if you think about it, this would yield you no information concerning the whereabouts of the only different coin. As we already know that one side will be heavier than the other. So that the following plan can be followed, let us number the coins from 1 to 12. For the first weighing, let us put on the left pan coins 1,2,3,4 and on the right pan coins 5,6,7,8. There are two possibilities. Either they balance, or they don't. If they balance, then the different coin is in the group 9,10,11,12. So for our second weighing we would put 1,2 in the left pan and 9,10 on the right. If these balance then the different coin is either 11 or 12. Weigh coin 1 against 11. If they balance, the different coin is number 12. If they do not balance, then 11 is the different coin. If 1,2 vs 9,10 do not balance, then the different coin is either 9 or 10. Again, weigh 1 against 9. If they balance, the different coin is number 10, otherwise it is number 9. That was the easy part. What if the first weighing 1,2,3,4 vs 5,6,7,8 does not balance? Then any one of these coins could be the different coin. Now, in order to proceed, we must keep track of which side is heavy for each of the following weighings. Suppose that 5,6,7,8 is the heavy side. We now weigh 1,5,6 against 2,7,8. If they balance, then the different coin is either 3 or 4. Weigh 4 against 9, a known good coin. If they balance then the different coin is 3, otherwise it is 4. Now, if 1,5,6 vs 2,7,8 does not balance, and 2,7,8 is the heavy side, then either 7 or 8 is a different, heavy coin, or 1 is a different, light coin. For the third weighing, weigh 7 against 8. Whichever side is heavy is the different coin. If they balance, then 1 is the different coin. Should the weighing of 1,5, 6 vs 2,7,8 show 1,5,6 to be the heavy side, then either 5 or 6 is a different heavy coin or 2 is a light different coin. Weigh 5 against 6. The heavier one is the different coin. If they balance, then 2 is a different light coin.

133. Train Journey

Mr. Grumper grumbles about bad time-keeping trains like everybody else. On one particular morning he was justified, though. The train left on time for the one hour journey and it arrived 5 minutes late. However, Mr. Grumper's watch showed it to be 3 minutes early, so he adjusted his watch by putting it forward 3 minutes. His watch kept time during the day, and on the return journey in the evening the train started on time, according to his watch, and arrived on time, according to the station clock. If the train traveled 25 percent faster on the return journey than it did on the morning journey, was the station clock fast or slow, and by how much?

Answer

The station clock is 3 minutes fast. If the morning journey took 65 minutes, and the evening journey therefore took 52 minutes, and the train arrived 57 minutes after it should have left, that is, 3 minutes early.

134. Basket of Eggs

A man is walking down a road with a basket of eggs. As he is walking he meets someone who buys one-half of his eggs plus one-half of an egg. He walks a little further and meets another person who buys one-half of his eggs plus one-half of an egg. After proceeding further he meets another person who buys one-half of his eggs plus one half an egg. At this point he has sold all of his eggs, and he never broke an egg. How many eggs did the man have to start with?

Answer

7 eggs. The first person bought one half of his eggs plus one half an egg ($3\frac{1}{2} + \frac{1}{2} = 4$ eggs) This left him 3 eggs. The second person bought one-half of his eggs plus one half an egg, ($1\frac{1}{2} + \frac{1}{2} = 2$ eggs) leaving the man 1 egg. The last person bought one-half of his eggs plus one-half an egg, ($\frac{1}{2} + \frac{1}{2} = 1$ egg) leaving no eggs.

135. Bottle of Medicine

A mother has three sick children. She has a 24-ounce bottle of medicine and needs to give each child eight ounces of the medicine. She is unable to get to the store and has only three clean containers, which measure 5, 11 and 13 ounces. The electricity is out and she has no way of heating water to wash the containers and doesn't want to spread germs. How can she divide the medicine to give each child an equal portion without having any two children drink from the same container?

Answer

Fill the 5 oz. and 11 oz. Containers from the 24 oz. container. This leaves 8 oz. in the 24 oz. bottle. Next empty the 11 oz. bottle by pouring the contents into the 13 oz. bottle. Fill the 13 oz. bottle from the 5 oz. container (with 2 oz.) and put the remaining 3 oz. in the 11 oz. bottle. This leaves the 5 oz. container empty. Now pour 5 oz. from the 13 oz. bottle into the 5 oz. bottle leaving 8 oz. in the 13 oz. bottle. Finally pour the 5 oz. bottle contents into the 11 oz. bottle giving 8 oz. in this container.

136. Numbers 1 to 9

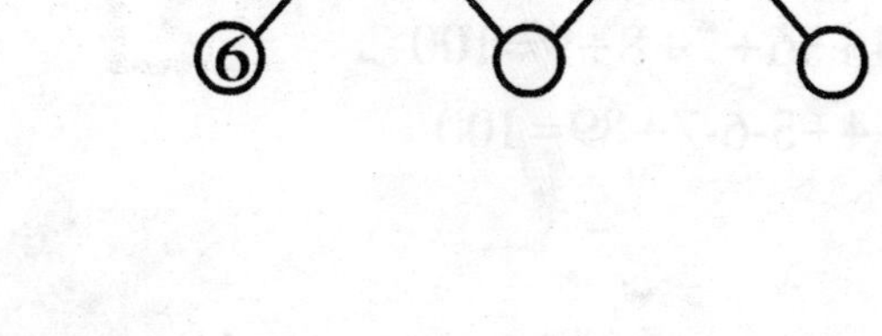

Can you arrange the numbers 1 to 9 in the circles so that each straight line of three numbers totals 18?

Answer

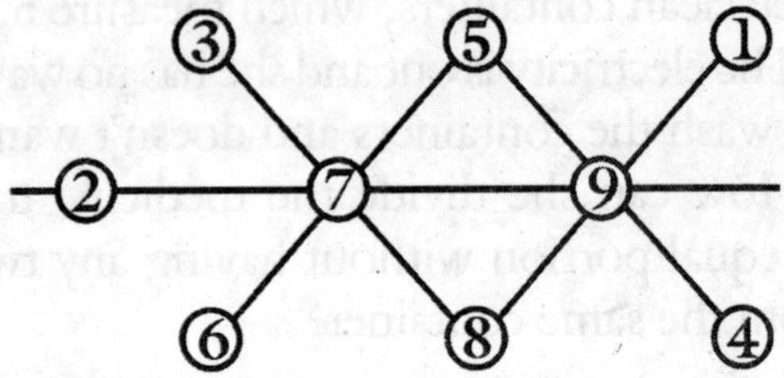

137. Make 100

Look at the following:

1 + 23 - 4 + 5 - 6 + 78 + 9 = 106

Notice that the digits 1 through 9 are used in order to arrive at 106. Using 1 through 9 in order, and using only addition or subtraction, create an equation that equals 100.

Answer

There are a lot of solutions for this riddle. Here are some of them:

12+3-4+5+67+8+9 = 100

123+45-67+8-9 = 100

1+2+3-4+5+6+78+9=100

1+2+34-5+67-8+9=100

1+23-4+5+6+78-9=100

1+23-4+56+7+8+9=100

12+3+4+5-6-7+89=100

12-3-4+5-6+7+89=100

123+4-5+67-89=100

123-4-5-6-7+8-9=100

123-45-67+89=100

138. Computer cord

The plug of the computer cord has 7 contacts located in a circle. It plugs into the outlet, which has 7 corresponding holes. Is there any way to number the contacts on the plug and the holes on the outlet, so that at least one of the contacts would fit in the corresponding hole any time you plug in the cord?

Answer

1526374 will match 1234567 always in one place. (each rotation brings the next point to a match.) The number is made by spacing 1234567 by twos, and then mod'ing everything by seven (i.e. 8mod7=1).

139. What row of numbers comes next in this series?

1

11

21

1211

111221

312211

13112221

Answer

1113213211 After the first line, each line describes the

previous line:
One One
Two Ones
One Two, One One
(and so on...)

CHAPTER 3

Brain Teasers

1. I am an odd number. Take away one letter and I become even. What number am I?

   Answer: Seven (take away the 's' and it becomes 'even').

2. Using only addition, how do you add eight 8's and get the number 1000?

   Answer: 888 + 88 + 8 + 8 + 8 = 1000

3. Sally is 54 years old and her mother is 80, how many years ago was Sally's mother three times her age?

   Answer: 41 years ago, when Sally was 13 and her mother was 39.

4. Which 3 numbers have the same answer whether they're added or multiplied together?

   Answer: 1, 2 and 3.

5. There is a basket containing 5 apples, how do you divide the apples among 5 children so that each child

has 1 apple while 1 apple remains in the basket?

Answer: 4 children get 1 apple each while the fifth child gets the basket with the remaining apple still in it.

6. There is a three digit number. The second digit is four times as big as the third digit, while the first digit is three less than the second digit. What is the number?

   Answer: 141

7. What word looks the same backwards and upside down?

   Answer: SWIMS

8. Two girls were born to the same mother, at the same time, on the same day, in the same month and in the same year and yet somehow they're not twins. Why not?

   Answer: Because there was a third girl, which makes them triplets!

9. A ship anchored in a port has a ladder which hangs over the side. The length of the ladder is 200cm, the distance between each rung in 20cm and the bottom rung touches the water. The tide rises at a rate of 10cm an hour. When will the water reach the fifth rung?

   Answer: The tide raises both the water and the boat so the water will never reach the fifth rung.

10. Can you find a quick and elegant way to add the numbers from 1 to 44 ?

    In other words, what is the sum you obtain in the

following case:

1 + 2 + 3 + ... + 44 =

Answer

Here's an elegant way:

1 + 44 = 45

2 + 43 = 45

3 + 42 = 45

....

....

22 + 23 = 45

Adding the above equations gives

1 + 2 + 3 + ... + 44 = 22 × 45 = 990

CHAPTER 4

Sudoku

Rules for Solving Sudoku

Solving a sudoku puzzle can be rather tricky, but the rules of the game are quite simple.

A sudoku puzzle is a grid of nine by nine squares or cells, that has been subdivided into nine subgrids or "regions" of three by three cells. See the following diagram:

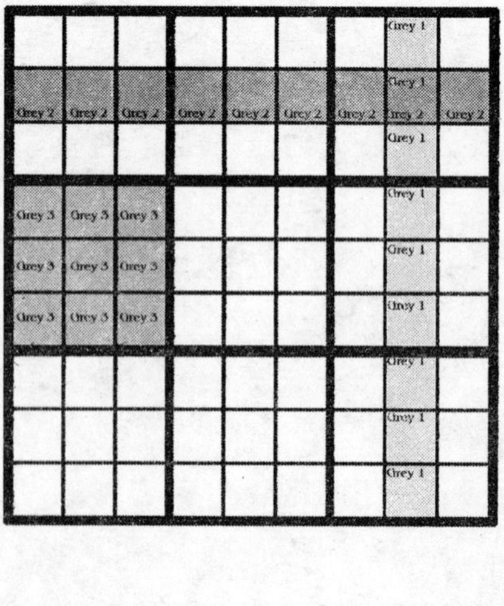

The objective of sudoku is to enter a digit from 1 through 9 in each cell, in such a way that:

Each horizontal row (shown in grey 1) contains each digit exactly once.

Each vertical column (shown in grey2) contains each digit exactly once.

Each subgrid or region (shown in grey 3) contains each digit exactly once.

This explains the name of the game; in Japanese, sudoku means something like "numbers singly".

Solving a sudoku puzzle does not require knowledge of mathematics; simple logic suffices. (Instead of digits, other symbols can be used, e.g. letters, as long as there are nine different symbols.)

In each sudoku puzzle, several digits have already been entered (the "givens"); these may not be changed.

The puzzler's job is to fill the remainder of the grid with digits –respecting, of course, the three constraints mentioned earlier.

A "good" sudoku puzzle has only one solution.

In spite of the game's apparent simplicity, sudoku can behighly addictive... While the first sudoku puzzle was published as early as 1979 (back then, it was called "Number Place"), the game's popularity really took off in 2005; it can now be found in many newspapers and magazines around the world.

Sudoku Puzzles

Place a number in the empty boxes in such a way that

each row across, each column down and each 9-box square contains all of the numbers from one to nine.

1.

			1	3				2
7								
2				4	9		8	
	5	4	6			7	2	
	1		3		8		6	
	7	3			5	8	9	
	2		5	1				9
								8
4				8	3			

Answer

5	8	9	1	3	6	4	7	2
7	4	1	8	5	2	9	3	6
2	3	6	7	4	9	1	8	5
8	5	4	6	9	1	7	2	3
9	1	2	3	7	8	5	6	4
6	7	3	4	2	5	8	9	1
3	2	8	5	1	7	6	4	9
1	9	7	2	6	4	3	5	8
4	6	5	9	8	3	2	1	7

2.

		4	7			3		1
								8
6	5	1			8			4
			9	4	6		3	
	8		3	7	1			
9			1			6	8	3
1								
7		2			9	4		

Answer

8	2	4	7	9	5	3	6	1
3	7	9	6	1	4	2	5	8
6	5	1	2	3	8	7	9	4
5	1	7	9	4	6	8	3	2
4	9	3	5	8	2	1	7	6
2	8	6	3	7	1	5	4	9
9	4	5	1	2	7	6	8	3
1	6	8	4	5	3	9	2	7
7	3	2	8	6	9	4	1	5

3.

			2				6	3
3					5	4		1
		1			3	9	8	
							9	
			5	3	8			
	3							
	2	6	3			5		
5		3	7					8
4	7				1			

Answer

8	5	4	2	1	9	7	6	3
3	9	7	8	6	5	4	2	1
2	6	1	4	7	3	9	8	5
7	8	5	1	2	6	3	9	4
6	4	9	5	3	8	1	7	2
1	3	2	9	4	7	8	5	6
9	2	6	3	8	4	5	1	7
5	1	3	7	9	2	6	4	8
4	7	8	6	5	1	2	3	9

4.

	1				4			
		6	8		5			1
5		3	7		1	9		
8		4			7			
			3			6		9
		1	5		8	2		4
6			4		3	1		
			2				5	

Answer

2	1	8	9	6	4	5	3	7
9	7	6	8	3	5	4	2	1
5	4	3	7	2	1	9	8	6
8	9	4	6	5	7	3	1	2
3	6	2	1	4	9	8	7	5
1	5	7	3	8	2	6	4	9
7	3	1	5	9	8	2	6	4
6	2	5	4	7	3	1	9	8
4	8	9	2	1	6	7	5	3

5.

1	3							
		2		5			3	
		9			2		8	
5			3				1	
			1		6			
	1				5			7
	9		4			3		
	8			2		5		
							6	4

Answer

1	3	8	9	6	4	7	2	5
7	6	2	8	5	1	4	3	9
4	5	9	7	3	2	1	8	6
5	4	6	3	7	9	2	1	8
8	2	7	1	4	6	9	5	3
9	1	3	2	8	5	6	4	7
6	9	5	4	1	8	3	7	2
3	8	4	6	2	7	5	9	1
2	7	1	5	9	3	8	6	4

6.

1			8	7	5	6		
					1	9	5	8
							1	
	2		7					6
			2	4	6			
4					3		7	
	9							
3	6	7	5					
		1	6	8	7			4

Answer

1	4	9	8	7	5	6	2	3
7	3	2	4	6	1	9	5	8
6	8	5	3	2	9	4	1	7
9	2	3	7	1	8	5	4	6
5	7	8	2	4	6	1	3	9
4	1	6	9	5	3	8	7	2
8	9	4	1	3	2	7	6	5
3	6	7	5	9	4	2	8	1
2	5	1	6	8	7	3	9	4

7

		2	6		4		9	3
	6			2		4		
5		4			7			
2		3						
		8				6		
						1		8
			3			7		5
		7		4			2	
8	2		9		6	3		

Answer

7	8	2	6	1	4	5	9	3
3	6	1	5	2	9	4	8	7
5	9	4	8	3	7	2	1	6
2	1	3	7	6	8	9	5	4
9	7	8	4	5	1	6	3	2
4	5	6	2	9	3	1	7	8
1	4	9	3	8	2	7	6	5
6	3	7	1	4	5	8	2	9
8	2	5	9	7	6	3	4	1

8.

3					8			
7		8	3	2				5
			9				1	
9					4		2	
				1				
	7		8					9
	5				3			
8				4	7	5		3
			5					6

Answer

3	1	9	4	5	8	7	6	2
7	6	8	3	2	1	9	4	5
5	4	2	9	7	6	3	1	8
9	8	5	7	3	4	6	2	1
6	3	4	2	1	9	8	5	7
2	7	1	8	6	5	4	3	9
1	5	7	6	9	3	2	8	4
8	2	6	1	4	7	5	9	3
4	9	3	5	8	2	1	7	6

9.

1				7			3	
8	3		6					
		2	9			6		8
6					4	9		7
	9						5	
3		7	5					4
2		3			9	1		
					2		4	3
	4			8				9

Answer

1	6	9	8	7	5	4	3	2
8	3	4	6	2	1	7	9	5
5	7	2	9	4	3	6	1	8
6	2	5	1	3	4	9	8	7
4	9	8	2	6	7	3	5	1
3	1	7	5	9	8	2	6	4
2	8	3	4	5	9	1	7	6
9	5	6	7	1	2	8	4	3
7	4	1	3	8	6	5	2	9

10.

9		8		5				7
7	2		1			8		
					6	3		
	9		3		7			5
				6				
4			9		1		6	
		5	7					
		1			9		3	2
2				3		4		1

Answer

9	3	8	2	5	4	6	1	7
7	2	6	1	9	3	8	5	4
5	1	4	8	7	6	3	2	9
6	9	2	3	8	7	1	4	5
1	8	7	4	6	5	2	9	3
4	5	3	9	2	1	7	6	8
3	4	5	7	1	2	9	8	6
8	7	1	6	4	9	5	3	2
2	6	9	5	3	8	4	7	1

11.

5	3		4					
6	9		8	3				
	1	8						
1	6	9		8				7
	2			7			6	
7				5		1	8	9
						9	5	
				4	8		3	1
					6		7	8

Answer

5	3	7	4	2	1	8	9	6
6	9	4	8	3	7	2	1	5
2	1	8	9	6	5	7	4	3
1	6	9	3	8	4	5	2	7
8	2	5	1	7	9	3	6	4
7	4	3	6	5	2	1	8	9
4	8	6	7	1	3	9	5	2
9	7	2	5	4	8	6	3	1
3	5	1	2	9	6	4	7	8

12.

			1			4	6	
1					5			
7		5		9	2	8	3	
8	7							
		3	8	1	6	7		
							5	8
	1	9	7	3		5		2
			6					3
	2	7			9			

Answer

9	3	2	1	8	7	4	6	5
1	8	4	3	6	5	2	9	7
7	6	5	4	9	2	8	3	1
8	7	6	2	5	4	3	1	9
5	9	3	8	1	6	7	2	4
2	4	1	9	7	3	6	5	8
6	1	9	7	3	8	5	4	2
4	5	8	6	2	1	9	7	3
3	2	7	5	4	9	1	8	6

13.

		9	8		6		3	4
4			1				8	7
2			4				9	
3								
		7		5		8		
								6
	7				2			8
8	4				1			3
6	3		5		8	7		

Answer

7	5	9	8	2	6	1	3	4
4	6	3	1	9	5	2	8	7
2	1	8	4	3	7	6	9	5
3	2	6	7	8	4	9	5	1
1	9	7	6	5	3	8	4	2
5	8	4	2	1	9	3	7	6
9	7	5	3	6	2	4	1	8
8	4	2	9	7	1	5	6	3
6	3	1	5	4	8	7	2	9

14.

	7				2			
		8				3		
	4	1			8		5	2
	8			7		9		
	5						3	
		7		1			4	
9	3		4			6	8	
		2				4		
			9				2	

Answer

5	7	3	6	4	2	1	9	8
2	9	8	7	5	1	3	6	4
6	4	1	3	9	8	7	5	2
4	8	6	2	7	3	9	1	5
1	5	9	8	6	4	2	3	7
3	2	7	5	1	9	8	4	6
9	3	5	4	2	7	6	8	1
8	6	2	1	3	5	4	7	9
7	1	4	9	8	6	5	2	3

15.

			2	6		7		1
6	8			7			9	
1	9				4	5		
8	2		1				4	
		4	6		2	9		
	5				3		2	8
		9	3				7	4
	4			5			3	6
7		3		1	8			

Answer

4	3	5	2	6	9	7	8	1
6	8	2	5	7	1	4	9	3
1	9	7	8	3	4	5	6	2
8	2	6	1	9	5	3	4	7
3	7	4	6	8	2	9	1	5
9	5	1	7	4	3	6	2	8
5	1	9	3	2	6	8	7	4
2	4	8	9	5	7	1	3	6
7	6	3	4	1	8	2	5	9

16.

1			4	8	9			6
7	3						4	
					1	2	9	5
		7	1	2		6		
5			7		3			8
		6		9	5	7		
9	1	4	6					
	2						3	7
8			5	1	2			4

Answer

1	5	2	4	8	9	3	7	6
7	3	9	2	5	6	8	4	1
4	6	8	3	7	1	2	9	5
3	8	7	1	2	4	6	5	9
5	9	1	7	6	3	4	2	8
2	4	6	8	9	5	7	1	3
9	1	4	6	3	7	5	8	2
6	2	5	9	4	8	1	3	7
8	7	3	5	1	2	9	6	4

17.

	2		6		8			
5	8				9	7		
				4				
3	7					5		
6								4
		8					1	3
				2				
		9	8				3	6
			3		6		9	

Answer

1	2	3	6	7	8	9	4	5
5	8	4	2	3	9	7	6	1
9	6	7	1	4	5	3	2	8
3	7	2	4	6	1	5	8	9
6	9	1	5	8	3	2	7	4
4	5	8	7	9	2	6	1	3
8	3	6	9	2	4	1	5	7
2	1	9	8	5	7	4	3	6
7	4	5	3	1	6	8	9	2

18.

	7			3				8
		2					1	4
1		8			4	7		
	8				5			
		7	8	1	2	9		
			4				8	
		5	6			3		9
7	2					6		
6				4			7	

Answer

5	7	4	1	3	6	2	9	8
3	6	2	9	8	7	5	1	4
1	9	8	5	2	4	7	6	3
9	8	1	7	6	5	4	3	2
4	3	7	8	1	2	9	5	6
2	5	6	4	9	3	1	8	7
8	4	5	6	7	1	3	2	9
7	2	9	3	5	8	6	4	1
6	1	3	2	4	9	8	7	5

19.

		7						8
				9		5		
		3	1		5	9	7	
	8			1	6		3	
	6		8	3			2	
	3	8	7		1	4		
		9		6				
1						8		

Answer

5	9	7	6	4	3	2	1	8
8	1	6	2	9	7	5	4	3
4	2	3	1	8	5	9	7	6
9	8	2	5	1	6	7	3	4
3	4	1	9	7	2	6	8	5
7	6	5	8	3	4	1	2	9
6	3	8	7	5	1	4	9	2
2	7	9	4	6	8	3	5	1
1	5	4	3	2	9	8	6	7

CHAPTER 5

Mathematics Tricks

Here are some tricks that help you to get faster results.

1. Addition of 5

When adding 5 to a digit greater than 5, it is easier to first subtract 5 and then add 10.

For example,

7 + 5 = 12

Also 7 - 5 = 2;

2 + 10 = 12

2. Subtraction of 5

When subtracting 5 from a number ending with a digit smaller than 5, it is easier to first add 5 and then subtract 10.

For example,

23 - 5 = 18

Also 23 + 5 = 28;

28 - 10 = 18

3. Division by 5

It is more convenient instead to multiply first by 2 and then divide by 10.

For example,

1375/5

= 2750/10

= 275

4. Multiplication by 5

It is more convenient instead of multiplying by 5 to multiply first by 10 and then divide by 2.

For example,

137×5

= 1370/2

= 685

5. Division/Multiplication by 4

Replace either with a repeated operation by 2.

For example,

124/4

= 62/2

= 31

Also,124×4

$= 248 \times 2$

$= 496$

6. Division/multiplication by 8

Replace either with a repeated operation by 2.

For example,

124×8

$= 248 \times 4$

$= 496 \times 2$

$= 992$

7. Division/Multiplication by 25

Use operations with 4 instead.

For example,

37×25

$= 3700/4$

$= 1850/2$

$= 925$

8. Division/multiplication by 125

Use operations with 8 instead.

For example,

37×125

$= 37000/8$

$= 18500/4$

$= 9250/2$

$= 4625$

9. Squaring two digit numbers

You should memorise the first 25 squares:

1	2	3	4	5	6	7	8	9	10	11	12	13	14
1	4	9	1 6	2 5	3 6	4 9	6 4	8 1	10 0	12 1	14 4	16 9	19 6

15	16	17	18	19	20	21	22	23	24	25
22 5	25 6	28 9	32 4	36 1	40 0	44 1	48 4	52 9	57 6	62 5

If you forgot an entry then do as indicated.

Say, you want a square of 13. Do this:

Add 3 (the last digit) to 13 (the number to be squared) to get $16 = 13 + 3$.

Square the last digit: $3^2 = 9$.

Append the result to the sum: 169.

As another example,

Find 14^2.

First, as before, add the last digit (4) to the number itself (14) to get $18 = 14 + 4$.

Next, again as before, square the last digit: $4^2 = 16$.

You'd like to append the result (16) to the sum (18) getting

1816 which is clearly too large, for, say, $14 < 20$ so that $14^2 < 20^2 = 400$. What you have to do is append 6 and carry 1 to the previous digit (8) making $14^2 = 196$.

10. Squares of numbers from 26 through 50

Let A be such a number.

Subtract 25 from A to get x.

Subtract x from 25 to get, say, a.

Then $A^2 = a^2 + 100x$.

For example, if $A = 26$, then $x = 1$ and $a = 24$.

Hence,

$26^2 = 24^2 + 100 = 676$

11. Squares of numbers from 51 through 99.

If A is between 50 and 100, then $A = 50 + x$. Compute $a = 50 - x$.

Then $A^2 = a^2 + 200x$. For example,

63^2

$= 37^2 + 200\times13$

$= 1369 + 2600$

$= 3969$

12. Any Square

Assume you want to find 87^2.

Find a simple number nearby a number whose square could be found relatively easy. In the case of 87 we take 90. To obtain 90, we need to add 3 to 87; so now let's subtract 3 from 87. We are getting 84. Finally,

87^2

$= 90\times84 + 3^2$

$= 7200 + 360 + 9$

$= 7569$

Another example, if we want to find 82^2. Then, we first take 80 ($\because$ its square is easy). Then add 2 to 82 which equals to 84. Then, $82^2 = 80 \times 84 + 2^2$

$= 6720 + 4$

$= 6724$

13. Squares Can Be Computed Sequentially.

In case A is a successor of a number with a known square, you find A by adding to the latter itself and then A.

For example, A = 111 is a successor of a = 110 whose square is 12100. Added to this 110 and then 111 to get A^2:

111^2

$= 110^2 + 110 + 111$

$= 12100 + 221$

$= 12321$

Note: This question can be solved by using the above rule also.

14. Squares of numbers that end with 5

A number that ends in 5 has the form $A = 10a + 5$, where a has one digit less than A.

To find the square A^2 of A, append 25 to the product $a\times(a + 1)$ of a with its successor.

For example, compute 115^2.

$115 = 11\times10 + 5$, so that $a = 11$.

First compute $11\times(11 + 1)$

$= 11\times12$

$= 132$ (since $3 = 1 + 2$)

Next, append 25 to the right of 132 to get 13225.

Product of $10a + b$ and $10a + c$ where $b + c = 10$.

Similar to the squaring of numbers that end with 5:

For example, compute 113×117, where $a = 11$, $b = 3$, and $c = 7$.

First compute $11\times(11 + 1)$

$= 11\times12$

$= 132$ (since $3 = 1 + 2$)

Next, append 21 ($= 3\times7$) to the right of 132 to get 13221.

15. Product of two one-digit numbers greater than 5.

This is a rule that helps remember a big part of the multiplication table.

Assume you forgot the product 7×9. Do this.

First find the excess of each of the multiples over 5: it's 2 for 7 ($7 - 5 = 2$) and 4 for 9 ($9 - 5 = 4$).

Add them up to get $6 = 2 + 4$.

Now find the complements of these two numbers to 5: it's 3 for 2 ($5 - 2 = 3$) and 1 for 4 ($5 - 4 = 1$). Remember their product $3 = 3\times1$.

Lastly, combine thus obtained two numbers (6 and 3) as $63 = 6\times10 + 3$.

16. Product of two 2-digit numbers.

The simplest case is when two numbers are not too far apart and their difference is even, for example, let one be 24 and the other 28.

Find their average: $(24 + 28)/2 = 26$ and half the difference $(28 - 24)/2 = 2$. Subtract the squares:

28×24

$= 26^2 - 2^2$

$= 676 - 4$

$= 672$

The ancient Babylonian used a similar approach. They calculated the sum and the difference of the two numbers, subtracted their squares and divided the result by four. For example,

33×32

$= (65^2 - 1^2)/4$

$= (4225 - 1)/4$

$= 4224/4$

$= 1056$

17. Product of numbers close to 100

Say, you have to multiply 94 and 98. Take their differences to 100: $100 - 94 = 6$ and $100 - 98 = 2$. Note that $94 - 2 = 98 - 6$ so that for the next step it is not important which one you use, but you'll need the result: 92. These will be the first two digits of the product.

The last two are just $2 \times 6 = 12$.

Therefore, $94 \times 98 = 9212$

18. Multiplying by 11

To multiply a 2-digit number by 11, take the sum of its digits. If it's a single digit number, just write it between the two digits. If the sum is 10 or more, do not forget to carry 1 over.

For example, $34 \times 11 = 374$ since $3 + 4 = 7$

$47 \times 11 = 517$

since $4 + 7 = 11$

19. Faster subtraction

Subtraction is often faster in two steps instead of one.

For example,

427 - 38

= (427 - 27) - (38 - 27)

= 400 - 11

= 389

A generic advice might be given as "First remove what's easy, next whatever remains". Another example:

1049 - 187

= 1000 - (187 - 49)

= 900 - 38

= 862

20. Faster addition

Addition is often faster in two steps instead of one.

For example,

487 + 38

= (487 + 13) + (38 - 13)

= 500 + 25

= 525

A generic advice might be given as "First add what's easy, next whatever remains".

Another example:

1049 + 187

= 1100 + (187 - 51)

= 1200 + 36

= 1236

21. Multiply, then subtract

When multiplying by 9, multiply by 10 instead, and then subtract the other number.

For example,

23×9

= 230 - 23

= 207

The same applies to other numbers near those for which multiplication is simplified:

23×51

$= 23 \times 50 + 23$

= 2300/2 + 23

= 1150 + 23

= 1173

87×48

= 87×50 - 87×2

= 8700/2 - 160 - 14

= 4350 - 160 - 14

= 4190 - 14

= 4176.

22. Multiplication by 9, 99, 999, etc.

There is another way to multiply fast by 9 that has an analogue for multiplication by 99, 999 and all such numbers. Let's start with the multiplication by 9.

To multiply a one digit number a by 9, first subtract 1 and form b = a - 1.

Next, subtract b from 9: c = 9 - b. Then just write b and c next to each other:

9a = bc

For example, find 6×9 (so that a = 6)

First subtract: 5 = 6 - 1

Subtract the second time: 4 = 9 – 5

Lastly, form the product 6×9 = 54

Similarly, for a 2-digit a:

bc = 100b + c

= 100(a - 1) + (99 - (a - 1))

$$= 100a - 100 + 100 - a$$

$$= 99a.$$

Do try the same derivation for a three digit number. As an example,

543×999

$= 1000 \times 542 + (999 - 542)$

$= 542457$

■ ■ ■